Behind More Binoculars

Behind More Binoculars

Interviews with Acclaimed Birdwatchers

Keith Betton and Mark Avery

PELAGIC PUBLISHING

Published by Pelagic Publishing
www.pelagicpublishing.com
PO Box 725, Exeter EX1 9QU, UK

Behind More Binoculars: Interviews with Acclaimed Birdwatchers

ISBN 978-1-78427-109-1 (Hbk)
ISBN 978-1-78427-110-7 (ePub)
ISBN 978-1-78427-111-4 (Mobi)
ISBN 978-1-78427-112-1 (PDF)

A catalogue record for this book is available from the British Library.

Cover image: *Puffins* by Robert Gillmor

Printed in India by Imprint Press

Contents

Preface

This is the second of our books of interviews with birders, birdwatchers and people interested in birds. When it came out a couple of years ago, *Behind the Binoculars* was well received and we had plenty more people on our list of potential interviewees, so we decided to produce another volume.

Producing a book like this is fun for the compilers – we meet interesting people, ask them impertinent questions and hear their stories. We hope you enjoy reading the interviews too.

One or other of us talked to each of the people in this book, and the interviews were transcribed, edited and then approved by the people we interviewed.

Spoken English is very different from written English, and we have tried to retain the conversational nature of the interviews but also make them relatively easy to read.

There are a few terms scattered through the interviews which might perplex some readers, and most of these are to do with the field sport of 'twitching'. Not all, in fact rather few, birders are twitchers, even though the popular media can't seem to get that fact straight. Twitchers are people who rush after rare birds hoping very much to see them. They twitch with a mixture of excitement and nervousness when they hear of a rare bird that they wish to see. If they fail to see it, then they have 'dipped', whereas if they see a bird they've never seen before it is a 'lifer' as it has been added to their 'life list' of birds (and by definition their 'year list' too). Many British birders have 'British lists', some have 'county lists', and others might have 'garden lists' – some have longer lists of lists.

Many birders do much of their regular birdwatching at a regular spot, often close to home, known as their 'patch'.

The order in which the interviews appear in this book bears no relation to the order in which they were done – or to anything else, except to what we thought was a good order.

When we paused for breath, with all the interviews completed, we saw that we had been lucky (maybe skilful?) in picking some fascinating people with a wide range of backgrounds and perspectives. We also realised that this book takes the reader behind the binoculars of famous birders and into their heads, their thoughts and emotions.

Each interview stands alone as an interesting account, but there are also some common themes, or differences, that leaped off the pages. We discuss these in the last chapter.

Keith Betton and Mark Avery
March 2017

Acknowledgements

We are grateful to Nigel Massen at Pelagic Publishing for encouraging us to put this book together, our interviewees for their openness and patience, and our partners (Esther Betton and Rosemary Cockerill) for their patience and help too.

Abbreviations

A-Levels	General Certificate of Education, Advanced Level Examination
AA	Automobile Association
AK-47	Avtomat Kalashnikova assault rifle
BB	*British Birds*
BBC	British Broadcasting Corporation
BBONT	Berkshire, Buckinghamshire and Oxfordshire Naturalists' Trust
BBRC	British Birds Rarities Committee
BNS	Bristol Naturalists' Society
BOC	Bristol Ornithological Club
BOU	British Ornithologists' Union
BOURC	British Ornithologists' Union Records Committee
BTO	British Trust for Ornithology
CAP	Common Agricultural Policy
CIA	Central intelligence Agency
CITES	Convention on International Trade in Endangered Species
DNA	deoxyribonucleic acid
DPhil	Doctor of Philosophy
EN	English Nature
EPC	extra-pair copulation
FoE	Friends of the Earth
FRS	Fellow of the Royal Society
G20	Group of Twenty of the largest national economies
GP	general practitioner
GPS	Global Positioning System

HSBC	Hong Kong and Shanghai Banking Corporation
ICBP	International Council for Bird Preservation
IUCN	International Union for Conservation of Nature
IWRB	International Waterfowl and Wetlands Research Bureau
ITV	Independent Television
JBRC	Junior Bird Recorders Club
JNCC	Joint Nature Conservation Committee
KPMG	a professional service and auditing company
M16	a US army rifle
MI5	Military Intelligence 5 (UK security service)
MP	Member of Parliament
MSc	Master of Science degree
NGO	non-governmental organisation
NHS	National Health Service
OBE	Order of the British Empire
OSME	Ornithological Society of the Middle East
PA	personal assistant
PhD	Doctor of Philosophy
RAF	Royal Air Force
RSPB	Royal Society for the Protection of Birds
SBSG	Sheffield Bird Study Group
SCC	Stop Climate Chaos
SOS	Sussex Ornithological Society
TRAFFIC	Trade Records Analysis of Flora and Fauna in Commerce
TTV	timed tetrad visit
UCNW	University College of North Wales
UN	United Nations
UNEP	United Nations Environment Programme
UV	ultraviolet
WCMC	World Conservation Monitoring Centre
WEA	Workers' Educational Association
WeBS	Wetland Bird Survey
WWF	World Wide Fund for Nature
YOC	Young Ornithologists' Club

FRANK GARDNER

*Frank Gardner is best known as the BBC's Security
Correspondent. He is an avid birdwatcher. While in
Saudi Arabia in 2004 he was shot six times in an attack
by terrorists, leaving him partly paralysed in the legs and
dependent on a wheelchair. He was born in the 1960s.*

INTERVIEWED BY KEITH BETTON

Where did it all start for you with birding?

I was about ten, my mother got me into it, and we thought she was
a bit mad standing on windswept North Sea beaches in Holland
when we lived there. I remember she was terribly excited one
winter when she saw a Nutcracker, which sounded quite fun to me
at the age of ten. But I couldn't get her fascination with it, as she
usually seemed to choose to birdwatch in very bleak and rather
inhospitable places.

But she persuaded me to give it a try, I was given my first pair
of fairly basic binoculars and it became quite fun. However, I was
completely put off it three years later when I went to boarding
school because the housemaster was a birdwatcher, and therefore
the epitome of uncool, so I wanted nothing to do with him and his
hobby and it took me twenty years to come back to it. And when
I think of all the amazing places I backpacked round in central
Asia, Latin America, Southeast Asia, Vietnam, Cambodia and
Brazil and I wasn't a birder then! I do remember going through a
forest climbing a volcano in Chile and seeing a wonderful big black

woodpecker with a red crest, and subsequently I realised it must have been a Magellanic Woodpecker.

It wasn't until I was living in Bahrain and my mum came to stay in spring. I took her out for a picnic and we sat in this lovely oasis and a bright yellow bird shot past, then a few minutes later a bright red bird went past. I realised it was time to find out what these were so I tracked down the natural history guide there – Howard King, who's still there today – and of course the yellow bird was a Golden Oriole on migration and the red one was a Madagascar Fody. People in Bahrain keep them as exotic pets and release them, so in Bahrain and Dubai you have these colonies of invasive birds like Zebra Waxbill, fodies and even African whydahs in some parts of Dubai. But what it meant to me was a need to find out what the shorebirds were, as they must be really exotic – only to find that they were plain old European Curlews, Whimbrels and Ringed Plovers – but I found that the tiny island of Bahrain had a very rich avifauna.

It was when I started travelling to Saudi for business that things really opened up. I went to the southwest where there is a wonderful fly route – you can see a lot of the African species in Saudi and Yemen such as Bateleur, Hamerkop, Dark Chanting Goshawk (if you're really lucky), Yemen Thrush, Olive Pigeon, and Amethyst Starling, which is the most stunning bird.

Was there anyone else in your life interested in birds?

Subsequently Dr Stuart Butchart who worked for Birdlife International in Cambridge. I've birded a few times with him in East Anglia and he showed me my first Stone-curlew in Suffolk and also my first Water Rail at Titchwell. He's also a wheelchair birder but it doesn't put him off and he goes everywhere.

What was your first pair of binoculars?

They were pretty crude, 7×25, heavy metal. The ones I have now are Steiner, 8×40. About four or five years ago, at my request, my mum gave me a scope and a tripod for my birthday, but to be honest I don't use it that much as the eyepiece is angled upwards, which is rubbish for me in a wheelchair as I need one that I can look straight down.

Were you a member of any bird clubs?

I did join the YOC but we were living in Holland at the time so I didn't get involved.

Did you use bird books or keep notebooks as a young birder?

I used Peterson, Mountfort and Hollom. I did keep very detailed notebooks when I was ten or so – and still have them – with maps of Hampshire where we lived, where the nests were, dates when they hatched. It's very interesting because the pattern of bird life has changed. My parents bought a cottage near Selborne in northeast Hampshire in 1971 – at the time the fields were wet and quite overgrown, and every time I walked through I'd put up Snipe, saw Woodcock there in the woods, Whinchat, and I saw my one and only Lesser Spotted Woodpecker in the garden – and have never seen them there since. Instead we regularly get things like Buzzard and Kingfisher.

Did you show any interest in Gilbert White?

We went to his museum, and in fact my parents' cottage was mentioned in his account of the Great Storm of 1774. I remember hearing Cuckoos in the 1970s, saw a few Tree Sparrows and House Sparrows, hardly any of which you see any more. Siskins were plentiful in the larches on the hangers.

It sounds to me as if you didn't get into twitching, you just enjoyed seeing things where you lived.

Between the ages of about ten and twelve I was quite scientific about it, plotting behaviour and movements, and I did a lot of drawing. Back then there was a thing called Rotring which was Indian ink drawing with a 0.3 mm fine pen and I produced a paper magazine of what was going on at the time which I'd give out to friends. I used to go and stay with friends from school in places like Norfolk and Romney Marsh and we'd go birdwatching together.

So did you start birdwatching post-university?

I didn't start again till I was thirty-one, triggered by the Golden Oriole/Fody experience, and realising I had enough time to go round and explore. I had a Mustang soft-top convertible so I was able to drive round Bahrain with the roof down and get down to a remote stretch of coast where I was stunned at the lines of Greater Flamingo there, so I bought a long lens and started to take photos of birds. Seeing European Roller on migration… and I was there for another two or three years and you could set the date by the days that the European Bee-eaters would come through in April and October. I'd literally wake up hearing their call and thinking, 'The Bee-eaters are here – it must be the 24th.'

Did you get into any tricky situations there?

It was always tricky in Saudi Arabia, particularly in the wild areas outside the city where the view is that a Westerner with binoculars has to be a spy – CIA, Mossad or whatever. But I used to go straight to the nearest National Guard checkpoint or police station and declare myself up front, where I was staying, and that I was there to look at birds, show them my bird book and ask if they knew where I could see them. So if I was stopped later on I could tell them to go and see Captain So-and-So because he knows what I'm doing. This would have been impossible without being able to speak Arabic, because they didn't have any English then. And there were some incredible birds – for example I've got a photo of six Steppe Eagles perched in a thorn bush. But definitely quite tricky, especially photographing with a long lens.

Is there anywhere in the Middle East you haven't been birding – for example Syria?

I have been to Syria – I spent an afternoon birding in Ghouta, which of course is now synonymous with the terrible gas attack in August 2013. It's quite high up in the foothills on the Syria side of the Lebanon Mountains and I was primarily looking for raptors, but saw Finsch's Wheatear there. I went to Palmyra in 1992 – this was just before I took up birding again, so of course I wasn't looking for Bald Ibis!

Probably one of the most fascinating places is Socotra, the island off the tip of the Horn of Africa, geographically part of the African continent but politically part of Yemen. I managed to get a visa and spent a brilliant week there as there are about a dozen endemics – including some which at the time were classified as subspecies such as the Socotra Buzzard, which is now a species on its own, Socotra Starling and Socotra Sunbird. I was staying right up in the mountains filming with a UN team of scientists who were trying to preserve it as a biosphere, and we saw Forbes-Watson's Swift. We went about 1,400 metres up into the Haghir Mountains staying in a remote village. I asked the locals about owls and they took me through a dark plantation and orchard and there were these two tiny Socotra Scops Owls – it's amazing somehow that you come across species when you least expect to find them.

Further north, in Saudi Arabia, there's a place called Al Baha in the southwest. It's pretty remote and bleak but in some of the desert wadis there are pine trees and in just one hour there I came across African Eagle-Owl, Brown Woodland Warbler, Arabian Waxbill and Long-billed Pipit.

Have you done any birding in Iraq?

I was there during the 2003 invasion staying in Basra and had a wonderful afternoon off. We pitched our tents on the roof of Saddam's old palace – well, I say old, but it was actually quite new, built for one of his reprobate sons, and I'm not sure it was ever lived in. There was a very big walled garden of several acres and it was a very benign time – there were White-tailed Plovers and what I suspect in retrospect were Iraq Babblers. This was April and there were some other fabulous species on migration.

I also went to Iran. It's really not a good place to have binoculars, but before I was injured I walked right up from the last bus stop in the north of Tehran to quite high up in the Alborz Mountains. That was the first and only time I've seen Red-fronted Serin.

I went to Qeshm Island and looked without success for Great Stone-curlew, which was disappointing, but apart from that there was nothing that I hadn't already seen in Dubai.

I've been to Afghanistan four or five times embedded with the military. At Kandahar airfield, at the western end of the camp, there's this dreadful cesspit just before the perimeter of the camp and (if you can stand the smell) last time I was there I saw flocks of Rosy Starling in full plumage.

In Pakistan I climbed from Islamabad into the Margalla Hills and saw White-capped Water Redstart.

In your travels you've obviously interviewed royalty and very senior politicians. Have you ever taken the opportunity to talk to any of them about birdwatching?

Only with the senior Bahraini ruling family, as I was trying to get them to protect some of the environment there. It's an uphill battle because generally in the Middle East they're not terribly environment-conscious. In Saudi Arabia, for example, so much of the habitat has been lost and you'd be lucky to see Dark Chanting Goshawk. And they've hunted so much wildlife to extinction – Houbara Bustard, which used to be all over the Arabian Peninsula, are very hard to find now. But the Emirates are getting better, they have done preservation exercises, but generally the Gulf ruling families like to hunt these things rather than watch them in places like Morocco, Pakistan and Iran.

When you were shot six times in Saudi Arabia in 2004 that clearly changed your life dramatically. You have managed to visit many great birding locations since then. Can you tell me about some of your more challenging expeditions in the last ten years or so?

Top of the tree would probably have to be the two trips to Papua New Guinea I made for a BBC documentary last year. Access-wise it was tough-going with a wheelchair, but much of the time in the lowlands we were travelling along the Sepik River. There were hundreds of Rufous Night Herons and Whistling Kites and the occasional Rainbow Bee-eater and Sulphur-crested Cockatoo – but otherwise disappointing. The Papua New Guinea Highlands, though, were superb – and not just for birds of paradise. We had a great local bird guide who knew his fruit-doves from his pigeons and we had a wonderful time camping in a grass hut up at 10,000 feet.

I've done two trips to Svalbard (the last piece of inhabited dry land between northern Norway and the North Polar ice cap), again for the BBC. One was in March when there weren't many birds around, just Snow Bunting and Rock Ptarmigan. But when we came in May it was brilliant, the frozen lakes were just thawing so the Barnacle and Pink-footed Geese were flying in, along with Purple Sandpiper and Red-throated Diver. Clearly visible just offshore were mixed rafts of King and Common Eider as well as Brünnich's Guillemot. To get around Svalbard I had to drive a snowmobile, and we had an armed guard with us at all times because of Polar Bears, which is compulsory by law.

In 2010 we went to Borneo as a family and I spent a cracking day cycling slowly down Mount Kinabalu from 8,000 feet up, using my trike adaptation, together with a local bird guide. I saw some wonderful endemics there.

Given what happened in 2004, do you ever fear for your life when you are working these days?

No, although we did visit the Saudi–Yemen border in late 2016 and spent time with a family that had suffered a rocket attack the day before.

Some bird reserves appear to be well-provisioned for people with reduced mobility. What more should be done in such locations to improve this further?

No turnstiles, please! These are a nightmare for anyone in a wheelchair. Also it would be really helpful if the doors to bird hides – and toilets – could have a pulling bar on the inside so we can pull it shut.

If we could pick somewhere in the Middle East you've never been, where would you go?

If it was peaceful and safe, Yemen, though I have been there before. Again, if it was safe, parts of Turkey, though there is one place that is safe, Mazandaran Province in Iran, and eastern Iran to see Pleske's Ground Jay, which is such a star bird.

Where's the best place in the world you've been to for birding?

Malaysia and Thailand have got to be high on the list. In Malaysia last year I went to Fraser's Hill, which was superb, though the visibility really isn't that good there, and in Thailand Khao Yai, which is a national park about two hours north of Bangkok. I also went to Khao Sam Roi Yot National Park, where you can see White-bellied Sea Eagle, plus Black-capped and Collared Kingfishers. You've got both jungle and sea there.

A place you wouldn't go back to again for birding?

I found New Zealand disappointing. It's quite hard work for such a big land mass. There are some lovely native species such as Tui, Kokako and Morepork, but so many of the birds are just boring imports from Britain. One great area is the Otago Peninsula east of Dunedin, which looks exactly like the coast of Devon, but the Albatross Observatory is great, with huge birds just flying in from the Pacific and landing – it's like being on an aircraft carrier. A little bit further on you've got Yellow-eyed Penguin, which to date is the only penguin I've seen in the wild.

The best and worst places you've been to as a traveller rather than a birder?

As a journalist I found Saudi Arabia fascinating, though it is pretty hard work. Once the people get to know you they're very hospitable, but the authorities can be very suspicious and quite unfriendly. Getting into the country itself can be quite difficult – there's no real culture of the authorities welcoming you in. It is starting to change very slowly, but the pervasive culture and belief is that you're not allowed to take photos or film – though there's no actual law that I'm aware of. I remember a couple of Japanese passengers from a cruise ship in Jeddah took a photo of a statue and were arrested, and I was arrested by the religious police for filming the melon market.

Yemen is right now in the middle of a civil war which hopefully will come to an end soon. There are kidnappings and everyone's got guns, but they're very friendly and not malevolent towards foreigners at all, quite the opposite, but you've got to keep your wits about you.

I skied in Lebanon during the tail end of the civil war with guys with M16s patrolling the ski queue, which was fun!

Going back to birds, do you have a particular bird family that you enjoy watching more than any other?

Sunbirds – they're so beautiful and colourful to look at and you've got them all the way from East Africa to Southeast Asia. The first one I saw was Purple Sunbird in Fujairah, which was within a month of resuming birdwatching – and seeing this glossy, iridescent purplish bird, then seeing the female and thinking it must be a different bird! Or Nile Valley Sunbird in southwest Saudi Arabia – which is almost a miniature bird of paradise.

How do you remember such detail?

It's a bit like how you remember where you were when you heard a song you like. I remember where I was when I saw a bird I enjoyed looking at. In Djibouti – went 1994, first time to sub-Saharan Africa and very thrilling to see African species like Speckled Pigeon, Yellow-billed Stork and Lesser Flamingo in Djibouti harbour with its bright red bloodshot eye. Arabian Golden Sparrow, Red-billed Firefinch – all real exotics, as everything I'd seen till then had been Middle Eastern.

If you could see a particular bird you haven't seen, what would that be?

A really good view of Andean Condor in the wild, and I still haven't seen White-tailed Eagle. And maybe Pleske's Ground Jay.

You must have met some other celebrity-type birdwatchers?

I went to David Attenborough's inspirational birds of paradise talk at the Royal Geographic and he signed a book for my daughters.

Andrew Parker, the Director-General of MI5, is supposed to be a birder, but when I asked him about this he seemed rather reticent, so maybe he is not so keen after all.

If you could meet someone, alive or dead, and just have half an hour with them, who would you choose?

I would like to meet Sultan Qaboos of Oman. He's been on the throne for forty-six years, one of the longest-serving heads of state in the world, and he's done a really good job with a country that's not very rich. It would have been interesting to meet Saddam Hussein.

Where in the world would you like to go birding with all expenses paid?

Peru. There would be many contenders for that slot, such as Antarctica, but Peru has such incredible avifauna diversity that it's right up there. Coastal desert, altiplano, montane and Amazon rainforest. Fabulous! Also, I've never been there.

What new bird would you most like to see?

There are quite a few star birds that have so far eluded me, even when I must have passed within a mile of them: Atlantic Puffin, White-tailed Eagle, Three-wattled Bellbird, Northern Cassowary and the kiwis of New Zealand South Island. But I think the one bird I'd most like to see in the wild – and preferably somewhere with stunning scenery like Greenland or Iceland – is Snowy Owl.

What is your favourite bird book?

Birds of the Middle East by Richard Porter and Simon Aspinall.

What is your favourite non-birding book?

I don't have one favourite book but I particularly enjoy anything by Robert Harris.

What is your favourite music?

Widor's *Toccata*.

What is your favourite film?

Apocalypse Now.

What is your favourite TV programme?

We've recently enjoyed watching *Billions* on Netflix, and before that we loved watching *Breaking Bad.*

ANN AND TIM CLEEVES

Tim Cleeves is a birder from Bristol who worked for the RSPB and Ann Cleeves is an author of detective stories. Both were born in the 1950s.

INTERVIEWED BY MARK AVERY

Tim, I came across your name when I was a teenager in Bristol and I was a member of the Bristol Ornithological Club which sent round a newsletter, in the post, of recent sightings every month, and I recall it being littered with birds with your initials next to them. And I'd never seen any of them! You must have been born in Bristol because you have, still, such a Bristol accent, unlike me, and it reminds me of home.

Tim: Yes, I was born in Hanham. And I started going birding locally, in Hanham Woods. Gordon Beakes was one of my mates and we'd go birding together.

Ann: …he's a lecturer in mycology at Newcastle University now…

Tim: …he actually passed all his exams! We were into natural history really and Gordon would be grovelling about on the ground looking at earthstars and things like that. I don't think we saw anything very notable as far as birds were concerned.

You've just put your hand on some notebooks.

Tim: Yes, I've brought out some of my earliest notebooks. I lost my first notebook but this is a transcript of Gordon's notebook and starts in January 1965, when I was thirteen. And we went to a couple of places called Vinny Green and Emersons Green – both are underneath the M4 now – and we saw our first Fieldfares and one glorious day we saw what was obviously a Chiffchaff. We had no idea what it was, except it was a warbler and we had to go and do some research on it.

I've got notebooks all the way from then, 1965, to today. Well, yesterday actually.

That's how it all started and we joined the junior branch of the Bristol Naturalists' Society (BNS) which had sections like Botany, Geology, Ornithology etc., and that was really important. Friends like Andy Davis, Dick Senior and Keith Vinicombe were all in the Junior BNS. Eventually the keener birders broke away from the BNS and formed the Bristol Ornithological Club in 1967.

When was the Pied-billed Grebe that forms the BOC logo?

Tim: It was 1962/63 at Blagdon Lake when Robin Prytherch and Harry Thornhill found it, and then it disappeared – it was the first for Europe. Robin drew the logo.

I was with my gran at Chew Valley Lake, because she knew someone who had a car, one day in August 1965 when the Pied-billed Grebe popped up again. It's here in my notebook for 30 August 1965. There was Bernard King, George Sweet, Mr and Mrs Lovell, Keith Fox, and 'the man who filmed the first Pied-billed Grebe', who was Harry Thornhill, but I didn't know his name then. And the bird was in Heron's Green Bay, although it was skulking, but it gave great views if you were patient.

Bernard King used to send us all postcards of bird records, and that's how you would know something was about – or had been about! Bernard was a bit of a mentor – he was one of the people who would take the time to help young birders. He was immensely important to us.

Ann: Whenever one of his young birders would get hitched he would take the woman aside. I remember him taking me aside, and saying that Timothy is a very skilful birder and it would be a terrible shame if you were to cramp his style. And he said the same to Kate when she went out with John Rossetti.

And what did you say?

Ann: I think I said I wouldn't lead Tim too far astray.

Tim: Bernard was a good bloke, as was Roy Curber, who was Bernard's wheels. Until he went down to Cornwall, when he had to drive himself, Bernard always seemed to get a lift in Roy's van.

You knew people like John Rossetti, who I was at school with, and Nick and Paul Andrew?

Ann: Yes, we know them still, John and Paul.

Tim: I know exactly when I met John. There were two Grey Phalaropes at the dam at Chew Valley Lake, in October 1968, and John and Paul and I met there.

First pair of binoculars?

Tim: A pair of Regent 8×30s and then a pair of ex-Navy Barr and Stroud 7×50s...

Ann:... and that's what you had when I first met you.

Tim: I sold the Barr and Strouds to Neil Thomson, the current skipper of the *Good Shepherd* on Fair Isle.

First bird book?

Tim: *Observer's Book of Birds* without a doubt.

I don't think I met you when we were both in Bristol, and I came across you years and years later when we both worked for the RSPB. I expect you did something in between?

Tim: When I left Rodway Technical High School in Mangotsfield I did taxidermy with a firm called Rowland Ward in London, who used to stuff anything from elephants to mice. It was a really weird place. I was in the 'Heads' department learning how to mount heads of Kudu, Lesser Kudu and Oryx. The guy I worked for was a complete bastard and he used to be drunk all the time, so he would have a row with someone each day after lunch. And you knew your time would come. I got sent downstairs to the 'Tusks' department where all the dregs ended up – a bunch of acid-heads (this was the 1970s after all). So I used to spend all day long, with a scraper and some white spirit, cleaning elephant tusks. I stuck that for about ten days and then I went back to Bristol and worked in what my friends called the 'Rat Factory' where we prepared biological specimens for schools. We would kill rats and fill them up with formalin and send them off to A-Level students to dissect.

Then I worked at a garage, on the petrol pumps, because I was saving money to go to India. I went out there with Paul Andrew and Roy Smith for five months. When I came back, I worked as the credit controller for a carpet company – chasing all their bad debts.

Are you good with a baseball bat then?

Tim: Pretty good, yeah. I can break a knee with no trouble. I did enjoy that job because the money was good and the craic was good. I was earning over £60 a week, which in 1975 wasn't bad money, and then I decided to do this job for the RSPB for £15 a week, living in a barn in central Wales, protecting Red Kites and Peregrines.

The Red Kites finished early and then you'd move on to Peregrines. At the time, the Peregrine was a rare bird – there were hardly any nesting in Wales. I lived in a caravan in Pembrokeshire. John Rossetti came and stayed with me over the bank holiday weekend. And the first day John was there, a guy was hanging down this cliff on a rope. So I said to John, go and get the police, and I went after the guy, who had run off. And he was running, and running and running. And I was chasing, and chasing and chasing. But eventually he stopped and I thought, 'I wonder what's going to happen now,' because I'd not done this before. And I said, 'Hello – what are you up to?' and he said, 'Birdwatching,' and I didn't argue with him and said, 'Cold isn't it? Do you want a cup of tea?' and he said, 'Yes, that'd be good.' So we went back to the caravan and I put the kettle on and then the police arrived. He got the maximum fine, I think, about £25 for wilful disturbance.

I went to Bardsey for six weeks as a volunteer and then applied for the warden's job at Hilbre Island, at the mouth of the River Dee. We were married by then: in 1977. We were there for four and a half years.

And then you eventually came to the RSPB?

Tim: I ended up in the RSPB, in the Central Region (now Midlands), based in Droitwich, with Carl Nicholson, and I was an assistant regional officer although there weren't a huge number of staff in a regional office at that time. I did a lot of work with our volunteer groups – such as starting the Symonds Yat Peregrine Watch, which involved liaison with the Forestry Commission.

When you look back, what's your feeling about working for the RSPB? You left ten years ago, didn't you?

Tim: I worked with some really great people. They weren't just workmates, they also shared a lot of the passions and concerns I had, and were really dedicated, and I learned a lot too.

Where do you go birdwatching these days?

Tim: We're very lucky. I go sea-watching at two main places – Newbiggin, which is cold and wet and horrible, but sticks out a long way, and Seaton Sluice. Off Seaton Sluice I've seen two really good birds: Bridled Tern and Ross's Gull, and we found both of them. But I missed the Fea's Petrel at Newbiggin.

What is your favourite bird?

Tim: I really like Med Gulls – they're really clean, and I like them in all ages. I love Peregrines because I've had great times with them. And Red Kites! I think I've got a Top Twenty, not a favourite.
Ann: And you like 'The Tubbies'.
Tim: The Tubbies, yes. They are Waxwing, Little Auk and Hawfinch – we christened them the Tubbies. They're really great, don't you think?

Tell us a bit about the Slender-billed Curlew.

Tim: Nightmare! It's become a nightmare.

It was 1998. In early May in Druridge Bay and what can you say? I saw this small curlew in a flock of Curlew, it landed and it had a very slender bill. And I took loads of notes on it but at that time I didn't have a camera with me. We saw it, Ann and me, on the Sunday and I had to get a train down to Sandy the next day so

I didn't see it again, but it was there until Thursday, I think. Lots of people were interested in it, and some got still photographs and some film of it. None of them were brilliant.

Ann: I don't think we claimed it at the time, did we? Because there were some local birders who said we hadn't told them about it and we said that we weren't sure exactly what it was.

Tim: But there were some positive things about 'the curlew', including involvement with the Slender-billed Curlew Working Group, and that was good, but it's fairly obvious that the bird is probably extinct. So if it's extinct, what about the bird we saw? Well, it's now no longer on the British List. And if you are going to claim what might be the last record in the world of an extinct (probably!) bird then you have to be sure – and I think the new information on the underwing pattern by Andrea Corso from Sicily means that we certainly can't be sure. So it's not on my British List because I only count species that are accepted by the BBRC.

Shall we move on? Ann, how did you come into Tim's life?

Ann: I first met Tim in 1975 when I was assistant cook in the bird observatory on Fair Isle, and he came as a visiting birdwatcher.

And why were you on Fair Isle? Because you were a bird-watcher or because you were a cook?

Ann: Neither! I had dropped out of university…

Why?

Ann: Because I wasn't really enjoying it. I was doing English literature and I thought I could read books anywhere, with the arrogance that you have when you are young. I was a peripatetic childcare officer – so, looking after kids in their own homes if their parents couldn't look after them. I was living in a bedsit in Putney and I went to the pub to say goodbye to a friend of a friend who was going off to be assistant warden at Fair Isle and he was moaning that it was going to be cold and miserable, and I said that I wouldn't mind, so he said that if I was serious they were desperate for an assistant cook. So I was on Fair Isle about a fortnight later.

Were you at all interested in birds at that stage?

Ann: No, I'm still not. I have no interest whatsoever. I'm quite interested in birdwatching – and the places where birds are.

So I was up on Fair Isle and it didn't really matter that I couldn't cook because the assistant cook's job was really to peel tatties and clean bathrooms – but I did learn. And I can still cook a couple of dozen scones standing on my head. And the next year I went back to be cook – so I can't have done too badly. I just loved the island, the place and the people, and had such a great time there. And then Tim came in the September. There weren't that many younger people turning up so it was quite nice.

And he came back the next summer after his contract with the Red Kites and the Peregrines, and worked on one of the crofts, and camped, and proposed to me there. So that was nice, although we didn't know each other very well.

Can you remember what birds he saw when he came to Fair Isle that first time?

Ann: Tennessee Warbler – that was the big one – two of them. And Upland Sandpiper too. But I've still seen a few things Tim hasn't seen in the UK: Siberian Rubythroat, American Kestrel, Bimaculated Lark.

I think I saw on your website that you started writing at Hilbre because there wasn't anything else to do.

Ann: At first I did go back to university – at Liverpool. So we were living this schizophrenic existence, the only people living on this tiny island, no mains water or electric, and then I'd walk across to the mainland at low tide and go off to be a probation officer in Liverpool. It was very weird.

I was very young – we both were. I was twenty-two and Tim twenty-six, and then when I got pregnant it was just too much to do all that tide-dodging. But there's not much to do there if you're not into birds – which I'm not. So I started writing then. My first novel was published in 1986, a little after we left the island, and I've written a novel a year ever since, so my thirtieth novel comes out in 2016.

How do you do that? Do you have another twenty novels planned?

Ann: Well, yes. I have lots still to write about. I'm just interested in people, so you want to explain how people come to be as they are. It's telling lies for a living – making up stories.

You're a successful writer.

Ann: Yes, but only in the last ten years really. In 2006 I wrote the first Shetland book and I am very grateful to Shetland for encouraging me to do that. And that won the Golden Dagger, which is like the Oscar of crime writing.

Do you think that was the best book you'd written up until then?

Ann: No, I'm not sure it was, but because of the Shetland background it captured people's imagination or the reviewers' imaginations. And I suppose it was a new way of doing traditional crime fiction.

My mum is a big fan of your novels – which is why I have her Christmas present with me for you to sign, please. She is keener on the Shetland ones than the Vera ones.

Ann: I think it is because of the Shetland setting – for people who have never been they enjoy it, and for people who have visited Shetland I think it reminds them of what a great place it is. And it gives us an excuse to go back to Shetland three or four times a year.

And Shetland is the only place where I out-sell Ian Rankin, so it's nice to know that the residents like them too.

What's it like having your book televised?

Ann: Of course it's different. But every reader reads a book differently – and creatively. A reader comes to the books with their own history and imagination and prejudice – so the pictures they see in their head aren't the same as the ones that I see in my head anyway.

I get on really well with the production team. I think it's a lovely story about how they came to be made. One of my books was found in an Oxfam shop in Crouch End by Elaine Collins, who was then a books executive with ITV. They were looking for a detective show to replace *Frost* on a Sunday nights and they were looking for something with a strong female lead. So she picked up one of my Vera books to read on the plane going on holiday. That's how that happened – and it all happened quite quickly, and rather informally, not through agents and things, because someone working for Elaine was my daughter's best friend at university so we had some connections right from the beginning. She came up and looked at the patch with the lead scriptwriter before we'd signed contracts, and it became a tradition that that would happen and we'd show them round and then at the end of the recce Tim cooks us all a curry.

And we do the same on Shetland too. We take the writers up and introduce them to people and show them places. Elaine produced both shows and now is creative director with the BBC. And she is married to Peter Capaldi, so Dr Who has sampled our curry too.

There's a strong sense of place in what you write, isn't there?

Ann: Yes. Shetland is very special but Northumberland is brilliant too – you have a wide variety of landscapes to work with, from Newcastle, which is really buzzy, and you have the bonny bits like around Hadrian's Wall and the beaches and the moors. But you've also got the post-industrial communities like the ex-pit villages and the places on the Tyne where they used to build ships. Part of the reason for wanting to show scriptwriters around is so that they know they have that really wide palette to use when they are writing.

Why crime?

Ann: It's always been my comfort area really – even when I was at university reading Shakespeare. I started off with Golden Age crime like Dorothy Sayers, Ngaio Marsh, Margery Allingham, those types of books. I read a lot of translated crime novels. You can learn so much of a culture's preoccupations by reading its popular fiction, I think. And it's a way of travelling vicariously too.

Although you say you aren't really interested in birds, there are quite a lot of birds in your novels.

Ann: They have played quite a big part in our married life. It's a sort of osmosis. If I see something, then I'll know it's something I've never seen before if it isn't.

Tim: I think you are being overly modest. You did point out an Arctic Warbler on Fair Isle once. And did point out that it had a single wing bar and a long, pale supercilium...

Ann: ...yes, but it was walking down the road about three feet in front of us. It wasn't that difficult really.

How do you write?

Ann: I'm an early-morning person. I'll get up before Tim and try to write 1,200 words a day if it's a first draft. Don't plot in advance. Don't plan in advance. I don't know where it's going to go at all when I start.

Really?

Ann: It's like telling a joke: you have to know how to pace it and where the tag lines come in.

Don't you know what the ending is when you start?

Ann: No. No idea.

Don't you know who is going to get murdered?

Ann: No. I don't know anything about it.

So you have to have a very strong view of the characters?

Ann: Yes, and you have to be prepared to go back and rewrite a lot. So the book I've just finished is fine – but I know I have to go back and rewrite all the scenes with one character because I haven't got them just right yet.

Do you think you might start another chain of novels: a non-Vera and non-Shetland one?

Ann: I'll certainly do another one or two Shetland ones but that'll probably be it. It's only a small place – I can't kill them all off.

Although you're not that into birds, do you have a favourite bird?

Ann: I like delicate waders – something like a phalarope.

What is your favourite non-bird book?

Ann: *Le Grand Meaulnes* by Alain-Fournier – it's a coming-of-age novel in quite a gentle way. And it starts with a great evocation of the French countryside.

Tim: *The Catcher in the Rye.* I came to it quite late and remember reading it on the clifftop in Pembrokeshire when wardening Peregrines. And it made a big impression on me – perhaps because I was on my own and it takes you back to adolescence.

Ann: Another coming-of-age book!

What about your favourite film?

Tim: We saw *Suffragette* the other day – that was very good but very sad.

Ann: Mine's *The Third Man* – I love Graham Greene

I love Graham Greene, but it wouldn't be my favourite Graham Greene book.

Ann: No, it's not my favourite of his books, but the film is wonderful.

Tim: I'm stuck in the 1960s – and I like *The Graduate* – another story about growing up!

What about favourite music?

Tim: I'm a big, big Bob Dylan fan. I've seen him eight times but he's rather lost me recently. We went to see Leonard Cohen and his songs are so evocative and rich – and not miserable at all even though lots of people think they are. *Closing Time* is one of the best riotous tales of pubs you'll hear.

Ann: I'd go for that as well. And I'd go for *Suzanne*.

Do you have a favourite TV programme?

Both: *Spiral* – it's a French detective series. Forget all the Scandies, *The Killing* and *The Bridge*.

ROY DENNIS

*Roy Dennis has overseen several schemes to reintroduce
wildlife to the UK, and has been heavily involved in bird
conservation for over fifty years. He was born in the 1940s.*

INTERVIEWED BY KEITH BETTON

At what stage did you start getting interested in birds?

At a very young age. I was born in Lyndhurst in 1940 – my mum had
to move from near Fareham to the New Forest to have me, because
our home was in a dangerous place in the war. We lived in a rural
area and I was allowed out all the time – building dens, catching
newts and collecting eggs by the age of eight. The big change came
when I joined the Boy Scouts when I was nearly eleven and wildlife
became much more about observing.

How old were you when you got your first binoculars?

To start with, I didn't have binoculars at all. Instead, I had an old
gunsight with a rubber eyepiece which I bought from an army
surplus store in Southampton when I was about twelve. It didn't
have great magnification, probably 6×, but I became really good
with it. I must have had binoculars by the time I was fourteen.

Were you encouraged by your parents?

They encouraged me to be a country boy and never bothered that
our little gang would wander off into the woods for the whole day.
I remember when it was nearly dark that our mums would shout

for us to come in and get washed for school in the morning. When I was small the army was camped in the orchard and woods and we used to go and have a look at what they were up to. So it was an idyllic childhood of just being allowed to get on with it, and all of us became really good little naturalists, because there's nothing like catching, keeping things and looking after them. I kept tame Jackdaws, one of which I taught to ride on the handlebars of my bike, and another time I reared three Shelduck which used to follow me around.

Did you join any local clubs?

The Scouts were the main thing as there weren't any wildlife clubs. My scout troop was really rural and their campgrounds were in a big wood near the Hamble River. To join it, you had to climb a rope to get into the hut up a monster beech tree, and that was a real test! My scout master was interested in birds and had a set of the *Handbook of British Birds*, so I knew that whenever I saw anything new I could go to their house and borrow that book – in the morning you could have tested me on Black-tailed Godwit or whatever I'd seen that day and I'd have soaked up everything the *Handbook of British Birds* had on that species. The next person who influenced me was Doc Suffern at Titchfield Haven, but also the Portsmouth group at Farlington Marshes – there was Dave Billett, Taff Rees, Eddie Wiseman and so on. Doc Suffern just took us young people under his wing and taught us.

My link with a group would then have been with Edwin Cohen and the Hampshire Ornithological Society. I never remember going to any meetings, it was a matter of writing letters and sending your notes in to Edwin and him either believing them or not. The limit was how far your bike would take you or you could hitch-hike, so my stamping ground was Warsash, the Hamble River, local woods, Titchfield Haven, Farlington Marshes, maybe Thorney, and the New Forest. Occasionally I went to Needs Ore, where I found a Richard's Pipit in 1961. In my mid/late teens I also started to go St Catherine's Point, probably with Eddie Wiseman and Billy Truckell, and we'd camp there for the weekend. We were allowed to climb all over the outside of the lighthouse catching birds – at night there were hordes of Whitethroats and warblers, Turtle Doves, and once we caught a Quail – it was fantastic. It was a brilliant place, we camped in an old pigsty no longer used by the lighthouse keepers, and that's also where I got into ringing. My trainer fell sick one day and had to leave me to watch the nets. I remember catching and ringing Firecrests, Blue Tits, Wrens, and then I got my licence. The Hampshire coast in the 1950s was probably the best training ground for birding in the UK.

You moved in 1958 to Lundy. How did that come about?

I went to Price's Grammar School where there were no biological choices, just pure and applied mathematics, physics and chemistry, or you did the arts. I visited Harwell for two or three days and the possible career plan for me was to do a sandwich scholarship in atomic engineering, but I found the place not to my liking. Suddenly I got this letter from Lundy asking me if I'd like to be an assistant warden for the summer and autumn, so I shot off there and never went to university.

We ringed seabirds, and my most exciting find was a Baltimore Oriole, which was a first for Britain. The warden and I caught and identified it (in those days the books weren't terribly good).

I left Lundy in November 1958 but by then I'd already got the promise of a job on Fair Isle and I went there in March 1959. I was the assistant with Peter Davis, which was just fantastic. Soon after arriving I found what I thought was a funny-looking bunting of some sort, so I rushed back to the Obs for Peter and we caught and identified it as a Song Sparrow. Then in September, when I was doing a trap round, I caught a Booted Warbler and Peter Davis did a fantastic detailed analysis of it in the bird room. It's now the first accepted record of Sykes's Warbler, but at that time they weren't split. Fair Isle's been a very important part of my life. I went there in 1959 and George Waterston arrived that September, which was the first time I met him, and I thought he was a brilliant guy. I regard him as one of the most important mentors in my life – he was so supportive and enthusiastic. He told me all about the Ospreys at Loch Garten, as 1959 was the first time they had a public hide and the first time they'd had young (though it was subsequently discovered they'd had young in 1954). He asked me to go and work for him in the RSPB in 1960, although he wouldn't take me on as warden, only as assistant warden, as he said I was too young to control the people who came to the Osprey Watch in those days – they were all lieutenant colonels, wing commanders and group captains. But it was an amazing opportunity to meet these people, many of whom had been through the war and were extremely interesting characters – Colonel Stanford, who wrote about wildlife; Guy Brownlow from Suffolk, who was a great birder. So I did the Ospreys for four summers – it was a great place to meet people. Fair Isle in 1959 and Strathspey in 1960 were the mecca for birdwatchers in Britain – you were either wealthy or in the services if you went abroad.

Fair Isle was where I got really interested in seabirds, which has remained a lifelong interest.

During the first two winters I worked with General Wainwright at Abberton Reservoir, catching ducks with Roy King. That was another really good opportunity to get to know wildfowl. I taught the General how to use mist nests. In another two autumns I got to know the Nature Conservancy people. It was a fantastic organisation – Max Nicholson at the head, Sir Arthur Duncan, Morton Boyd, Joe Eggeling. Morton was another good mentor for me, and I went with him twice to the seal islands off the Uists and to North Rona where we were branding, sexing and tagging Grey Seal pups – and we saw plenty of birds as well. Estlin Waters came with us on the second trip.

In the early 1960s were you ringing the Ospreys yourself?

We weren't allowed to ring them at that time – it was felt that it would be too damaging, they were far too shy. We had some unusual views on Ospreys then, that they would desert, but when you had only one pair and a lot of people looking at them, there was no room for error. Doug Weir ringed the first in 1968.

How did the opportunity arise to go back to Fair Isle?

In 1963 Peter and Angela Davis decided to leave Fair Isle and Peter recommended me to George to take over the bird observatory. I was ready for a move and George thought that I and my new wife Marina would be a great couple to run the observatory, for it's a joint post. We were there from 1963 to 1970 and had our three children at Fair Isle. Since that time I've been a director of the Fair Isle Bird Observatory Trust, then chairman, and now I'm an honorary president. It was just a marvellous place to live and work, we made friends that have lasted a lifetime, and there's no better birdwatching than Fair Isle.

You were about thirty when you left Fair Isle. Had you done some overseas birding by that time?

Not really. When I was on Fair Isle I did a winter tour of Holland for a Dutch/UK friendship group, giving talks about Fair Isle Bird Observatory. I went round the Netherlands meeting ornithologists from various places and was taken to Arnhem and some of the goose sites.

So your RSPB career came along – was that down to George Waterston again?

George saw that things were getting busier in bird protection in Scotland and there wasn't time for him to keep coming up from Edinburgh, so it was decided that I'd get a job as Speyside representative and I went to live in Strathspey. Soon, though, it was very clear to me that there was a bigger job in the Highlands as a whole rather than just at Loch Garten, because North Sea oil started in 1972 and there were major industrial developments taking place around the Cromarty Firth, building oil rigs, the aluminium smelter and so on. I went to The Lodge and the really key person was David Lea, the RSPB Deputy Director at the time – I liked David, and he and I basically wrote my job description. I was to be Highland Officer, covering the whole of the Highlands, which was a much more fulfilling job than just being involved in Strathspey on rare birds. It meant dealing with nature conservation alongside planning and working with the different county councils, and very quickly I was thrown into all the business about protecting the Moray and Cromarty Firths from a huge number of developments to build oil platforms, a refinery and proposals for a massive rig-building yard in Dunnet Bay in Caithness. My equivalent at Nature Conservancy, Andrew Currie, and I did a lot of really high-profile stuff at that time. And then, of course, there were some superb birds starting to breed in the Highlands, such as Goldeneye, Lapland Bunting and Temminck's Stint, and great birders who were my field assistants. And the people of the Scottish Highlands and Islands were marvellous.

Was your time taken up by trying to track down egg collectors and make sure nests were protected?

Egg collecting was a major problem in those early days. Weekends from March to June were just full on, because you knew the collectors would come up on a Friday. We made some good catches but it was a lot of hard work and very disheartening when you'd found a new eagle nest and you'd go back and it had been robbed. By the mid-1970s there was great pressure in stealing birds of prey for falconry – either chicks or eggs to hatch. I think we lost one-third of our Peregrines to robberies in one year in the late 1970s – and at that stage Peregrines were rare because of the pesticide era. We did all sorts of things, such as painting their toenails with fluorescent paint which we could see under UV light. I remember going to RAF Lossiemouth and saying that I believed the Peregrines they had in their bird scaring unit were stolen – they wouldn't let me on the base until I told a senior officer that I'd get the police! We shone the light and proved the birds had been purchased in Cumberland. So there were some really good cases. The government's Advisory Committee for Birds in Britain, which I belonged to for a long time, set up a major working party on bird of prey law so things were being tightened up all the time and robberies became less common.

You had the Highland Officer job until 1986, when you became the Regional Officer for Northern Scotland – how different was that job?

Not that different. It was probably some fairness coming from The Lodge. I was doing much the same as the other regional officers but I was called the Highland Officer and presumably paid much less!

The RSPB had bought the area round Loch Garten, then the main Abernethy Forest came up – there was enthusiasm in Scotland but there were worries at The Lodge about the huge size and cost of it. Ian Prestt was going to make a crunch visit and it came down to me to persuade him – which I did, and he went away convinced about the purchase, which was tremendous. It was then my task to go to the RSPB Members Conference and talk about why we needed to kill so many deer.

Getting Abernethy linked in with my thinking about large-scale ecological restoration, and quite a chunk of my work in the 1980s when I was with the RSPB was moving the society to think about large ecosystems and ecological restoration. It really got up my nose when people would say, 'Oh, look at these beautiful great hills!' and I'd say, 'No, they're knackered, the woodland's been destroyed by overgrazing and burning, and we can't keep on having heather everywhere!' So it was a really interesting time.

We also had the excitement of seeing Osprey populations growing fast, and the reintroductions of White-tailed Eagle (which I had helped start on Fair Isle in 1968) and then Red Kites to Scotland and England. These were all extremely positive and successful projects.

I carried on till 1990 but I could see that the RSPB was changing, the number of people in my office on the Black Isle was increasing, and my opportunities to go into the field were getting fewer. My job was getting more to do with budgets, meetings and plans and it wasn't what I wanted to do with my life. I believed that being in the field, meeting people, doing good wildlife work, being proactive with projects, persuading landowners and industrialists and being able to appear on TV and radio at the drop of a hat was more important than being in an office. And nowadays people in my equivalent position have far less visibility in the countryside. In some ways the RSPB, and the Nature Conservancy, were in their heyday in the 1980s.

I always wanted to do positive projects and I realised I could do more if I left. I took a chance and went independent, and I never thought I'd be soon sitting on the main board of Scottish National Heritage, the Cairngorm working party and the Red Deer Commission. That was really exciting. If I did it again, I'd probably leave five years earlier, as I found that I just didn't fit into an organisation that was becoming more and more bureaucratic.

I decided first of all to do consulting work, including going at the request of the RSPB to the Arabian Gulf to help deal with oiled birds and the oil spills as the Gulf War ended.

When did you set up the Highland Foundation for Wildlife?

When I left the RSPB and became a consultant, there was tremendous pressure to get bigger and have staff – so the big problem was trying to keep small. I realised that if I wanted charitable funding for projects I needed to set up a charitable organisation. A lawyer friend in Edinburgh helped me to start the Highland Foundation for Wildlife in 1994, aimed very much at species reintroductions and ecological restoration.

I would say I've been lucky in my life, but I've worked damned hard – and one of the breaks I had was in my last year with the RSPB. Driving home from Inverness I saw a German car on the roadside near a Peregrine nest. I got my scope out and could see people on the cliff, so I called the police in Inverness. They went to one end of the dual carriageway, I went to the other, and we caught them. They'd been at that nest but we couldn't find any eggs. We took them to Inverness and after a bit of questioning they refused to speak English – an interpreter was brought in but they then spoke a slang German. In the morning the Sheriff felt sorry for them and let them go. Their car was stopped at Dover and under the dashboard was an incubator with Peregrine eggs. The RSPB at The Lodge found out that there was a law which allowed a criminal prosecution to be brought in Germany, so it meant that I and the policeman had to appear in court in Mannheim. The guys got hammered by their courts so that was a good result.

Next day I set off on an adventure which changed my visions for wildlife. I'd read about a man who was researching Lynx and Wolf and I drove to see him in the Alps in southern Germany. There I met Christoph Promberger, studying Wolves in Romania, his professor Wolf Schröder, who was studying Brown Bears, and a colleague who was studying Lynx – we became great friends. I visited Romania with them, we went to the Yukon, and they invited me to become a member of the newly founded Large Carnivore Initiative for Europe, so I went to their meetings. The Initiative was set up by Magnus Sylven, who was working for WWF International in Switzerland. He was the man that helped me collect the first Red Kites from his PhD study area in southern Sweden for the reintroduction project started in 1989. So the involvement with trying to get Beaver and Lynx back to Scotland became a big part of my life. One is now with us, and I still long to see the Lynx back home.

How do you spend most of your time?

There's a lot of work in our projects on Ospreys at home and abroad, as well as reintroduction plans for other species of birds and mammals, including three successful translocations of Red Squirrels in the Scottish Highlands. I do quite a lot of work advising estates and others on ecological restoration and wildlife conservation, I get requests to attend meetings and give talks, and I give quite a lot of time free to good causes in nature. I'm starting to sort out my years of correspondence, field notebooks and diaries with the aim of writing more books. And we've just done a big renovation of our house, much of it ourselves, and it's great fun having a young daughter starting to keep tadpoles and follow other country pursuits.

Of all the places you've been to, which has impressed you the most?

I really enjoy Scandinavia – in fact northern lands and deserts appeal to me more than tropical ecosystems. My older daughter works on tropical forest conservation in Southeast Asia and I've found trips with her fascinating, but I'm not sure I'd like to work there – it's too difficult for fieldwork.

My first trip to Mongolia and Siberia was special. It was for Mark Beaman. He was at Aberdeen University, said he'd be setting up a tour company when he left university and that he'd love me to be one of his leaders on the first trip in 1980. He took the first group and I took the second group – that was exceptional. I'm not a twitcher, so when I go abroad I'm as interested in how the ecosystems and conservation are working, how the people live, how they farm, what they eat. My latest trip was to the Sichuan Mountains of China where I went to look at Giant Panda conservation, which was fascinating. I'd actually wondered if the Giant Pandas were a bit of a waste of time, costing $1 million a year and proving difficult to breed. Iain Valentine, in charge of Giant Pandas at Edinburgh Zoo, invited me to go with him – he must have told them I was someone special as they took us into areas he'd never been in. The area they took us to was twice the size of the Cairngorms National Park and three times the size of Yosemite, and has this incredible montane ecosystem with 4,000 species of trees and plants, Clouded Leopard at the bottom and Snow Leopard at the top, and over 400 species of birds. This incredible protected natural ecosystem is all based on Giant Panda protection, and I think the real story is not about captive-breeding Giant Pandas but the ecosystem which the Chinese Forest Service is protecting for the nearly 2,000 wild pandas.

When you look back on your successes, what do you think?

In particular I'm really pleased that Ospreys, Red Kites and White-tailed Eagles have done so well and that we do have Beavers in this country. It's been a hard slog. And it's been great to help with restoring Ospreys in Spain, Portugal, Italy and Switzerland. It's been fun to have been involved since the beginning, but there is still so much more to do. I like that we defeated the huge exotic conifer onslaught of Sitka Spruce and Lodgepole Pine in the Flow Country and that our views on ecological restoration – now often called rewilding – are becoming mainstream.

I find it extremely disappointing that our bureaucracy often prevents proactive projects – there are too many over-cautious scientists and risk-averse bureaucrats. For example, the Curlew went on the Red List, so immediately the RSPB and BTO said they needed money for more research. I'd have thought there was enough research on Curlew to know what to do – but it's much harder to change how people use the land than to do research. Research takes three years, you get the money and you produce a report, so you've been successful – but you haven't always solved the problem. In general, world wildlife is declining even as the numbers of people working in nature increases, so we've got to be far more entrepreneurial, far more hands-on, and make things happen in the field. The worst thing is never to have tried. We need fewer people behind desks and more boots on the ground, taking risks. And nature must get a much higher political profile and, for long-term security of the earth, at least 40% of our land and oceans must be managed as natural ecosystems for nature and its ecosystem services. I wish we had tried to bring home the Lynx in the 1980s – it would be here by now.

You mentioned various people who interested and inspired you. What about people you've never met but would have liked to meet?

I'd love to have met Aldo Leopold, who wrote *A Sand County Almanac* – he was the first real wildlife ecologist and sadly died in a fire. I always suggest his book whenever people ask me what book they should read if they're interested in ecology. Pre-eminent on the world stage, in my opinion, are the views and writings of Edward Wilson, Paul Ehrlich and Jared Diamond from the United States, and Tim Flannery in Australia.

If you had to choose a bird family that you particularly like, what would that be?

People always ask what my favourite bird is but I've never felt the need for a favourite – they are all special.

If we could give you an airline ticket to anywhere in the world you haven't been, where would you go?

I'd love to see Tigers, and also Emperor Penguins – not together!

If you could think of someone in history you could have met, who would you go for?

Probably a benign dictator who could have sorted nature conservation out very quickly.

What bird book would you take to a desert island?

A Sand County Almanac by Aldo Leopold.

And what about a book not connected with conservation?

I read a very eclectic selection of books from the Forres library – fiction and non-fiction. I enjoyed the Henning Mankell series, for the principal reason that the action took place where I collected the Red Kites in Sweden.

What do you watch on TV?

I enjoy good documentaries. I love a good game of football as long as it features one of the top teams. And I love rugby union (though if Scotland start to lose to England I lose a bit of interest).

What music do you like?

I like Scottish folk music, especially the great musicians of Fair Isle. There was nothing more fun than Bobby Tulloch playing and singing at a dance on Fair Isle long ago.

KEVIN PARR

Kevin Parr is a writer of a novel on twitchers but also articles and books on fishing, who was born in the 1970s.

INTERVIEWED BY MARK AVERY

What birds do you see when you look out of the window over farmland in Dorset?

This view is great, and I like the fact that it is limited by the hilltop up there – so anything you see is certainly within binocular range. We see quite a few birds of prey: Buzzards and Sparrowhawks nest within sight of the cottage and Peregrines nest a mile and a half away on a pylon so we see them sometimes too, and we've seen Hen Harrier, Merlin (which I think was nobbled by the Sparrowhawk), Goshawk, Red Kite – and Kestrel of course. If there's a westerly wind it pushes up the hill and creates some lift and the raptors come and sit on it – and Ravens play in it too.

Just outside here, on the feeders, we get the usual birds but also Reed Bunting and occasionally Marsh Tits.

But the highlight in the garden was a Wryneck – the one and only Wryneck I've ever seen. I'd actually walked around to look at the colony of Grass Snakes we have on the compost heap and a small bird flew up, and it was one of those strange moments, where I knew straight away it was a Wryneck even though I'd never seen one before. We had brilliant views; it was incredibly confiding. It flew over and landed on the garden fence. That has been the highlight.

Where did you grow up?

I grew up near Winchester in a tiny hamlet – so, a rural environment. The outdoors was very much my playground from an early age. We'd be out and about, playing in the straw and hiding from farmers who didn't like us playing in their straw bales. I know it's dangerous but it was fun!

How did you get into birding?

There was no epiphany – I just remember always being interested in birds. Mum tells me that when I was young, preschool, we were walking home after seeing my older brother off to school, we were walking up the little lane by the house, and there was a Robin singing, and she says I just stood there for ten minutes watching it and listening to its song. So it started early and my mum and dad were happy to encourage it. I joined the YOC and a local birding group. It isn't just birds for me, it's all wildlife, but birds are special.

Was school important to you in encouraging an interest in wildlife?

No, largely irrelevant. My primary school was a tiny country school so liking birds wasn't a bad thing. When I went to secondary school I didn't mention it much. It was more important to get cool jackets from charity shops, waistcoats and floral shirts than to talk about birds.

How would you describe yourself as a birder? Do you go to Portland?

I would describe myself as an opportunist birder. We've lived in Dorset for four and a half years and I've still not got over to Portland…

Well, if the Wrynecks are coming to you…

Yes, that's true! The nearest bit of the coast to here is West Bexington, which is very well watched and has turned up Black Stork and Little Bustard recently. But if I go to the shops or go anywhere I take my binoculars. I don't go specifically birdwatching but I am always birdwatching. You'll notice that as we talk I'll be glancing out of the window all the time.

And there are things like going to listen to Nightingales and watching the Hobbies catch dragonflies over the heath that I do every year as part of the routine of the year.

What binoculars do you use?

I've got some Olympus 10×50 which I've had for about ten years. I started with some 8×30 by Swift.

Was there a bird book that was important to you when you were young?

The book we had at home was the *AA/Reader's Digest Book of British Birds*, and that was the family bird book. But the book I read the most was *Birds of Prey* by Glenys and Derek Lloyd, I think by Hamlyn, and it was perfect for a kid. It was full of facts and figures and I remember that the claw of a Harpy Eagle is as big as the claw of a Kodiak Bear. I remember it had Andean Condors in the book and it said, quite correctly, that Andean Condors would come down to the coast to feed on whales (dead ones) – but I got that a bit wrong and told everyone at school that Andean Condors were so big they could pick up whales and take them back into the mountains.

I came across your name when I bought your book, *The Twitch*, which I really enjoyed and reviewed on my blog – in fact it was my blog's 'Book of 2014'. The book is basically about a bunch of birders trying to see the most birds in a year and one of them cheats, by bumping off some of his competitors.

That came about when, around ten years ago, a White-tailed Eagle turned up at Cholderton and we went to see it after lunch in the pub. I've seen lots of White-tailed Eagles but to see one by the A303 seemed ridiculous and worth doing. The farmer was charging £5 for people to park. There were half a dozen other people there too. It was chatting to them that got me started. One guy had done 350–400 miles to see three birds that day, and this got me thinking that these people are fanatical. What I wanted to do, originally, was to join in and write a 'year as a twitcher' type book about the people, the places and the stories behind the people. And we went down to Zennor in Cornwall when a Snowy Owl turned up which I went to see and these three birders appeared, with telescopes and the works, saw the owl and were gone in a matter of moments off to the next bird. What's that about? There would be plenty to write about. But about the same time the BBC made that documentary of *A Very British Obsession* which really covered the subject and I couldn't get any publisher interested.

But the idea stayed in my brain, and I really wanted to write about it, and I gradually thought about writing a novel about a guy who is a real stickler for the rules and gets miffed if anyone claims a bird they might not have seen but is quite happy to push someone off a cliff to win a bird race. And it was such good fun to write.

You published it with a publisher called Unbound which is a sort of crowdfunded deal. Can you explain that, please?

You pitch the idea and if Unbound like the idea they offer you a platform to crowdfund for it to cover the production and marketing costs. Then you are relying on the public to buy your idea. So friends and family give it an initial surge and then you have to work to get people to buy the idea.

So what do they do?

I think I had to raise a five-figure sum for a run of about 1,500 books. Ultimately, once – if – the book gets into profit, further down the line, the author's cut is 50% of the profits.

The pledgers get little perks depending on how much they contribute, like being invited to the launch party, and there was a 'Birding with Kev day', and they all get the book and their name in the back of the book.

I loved *The Twitch*, and I hope lots of people will buy it and read it, but it was only after I had read it that I realised you had emailed me quite a while back for a quote about Cormorants

for an article you were writing about fishing. And that is your 'real' writing life, isn't it? And you are a fisherman too.

I am largely a coarse fisherman, which means that I am a freshwater fisherman who doesn't fish for game fish (basically Salmon and trout) with a fly rod. I do a bit of fly-fishing and sea-fishing but it's largely rivers and lakes for me. The location is important for me, and when I was working in an office in a stressful environment, when the weekend came round I had to go to a really nice place to fish to get the best feeling out of it.

I used to fish the River Kennet a lot. To fish there on a nice day is a great way to lose yourself.

So are you going to tell me that it didn't really matter whether you caught a fish or not?

No! It mattered a lot. I always wanted something tangible to get me through the week ahead, so not catching a fish would be a bit like dipping on a twitch for a twitcher – except much more important to me. You'd feel there was something missing if you didn't catch a fish. I'm more relaxed about it now that I have a less stressful lifestyle. Nowadays, I have more time, and catching the fish is a bit less important, but catching fish is why I am there ultimately.

The biggest thing about fishing, though, is not catching the fish but connecting. There's something about water, because it is such an alien environment to us. You've generally got an environment that is still fairly self-sufficient, and a habitat that is sustaining, but it's also something that we can't relate to because we don't like getting wet unless we have a towel and can dry ourselves. And we can't exist in that world because we'd drown – and so the only way to make a connection is with a rod and a line, and that gives you a purpose. And you suddenly feel part of everything around you, and you feel involved. Anyone can go and sit by a river, but without the fishing I always feel I could be anywhere. The fishing connects you to the place.

And everything starts to carry on around you. I see lots of birds because, as you sit there, you become part of the scene. That's what it's about. Being involved and being part of the natural world –

even though I do want to extract something from it. And I do want to lovingly put it back. It is a contradiction in so many ways.

The urge is definitely from a hunting element.

This year how many days will you have gone fishing?

Not many. Probably about a dozen, but I've had a busy year writing a book and I've been a bit strapped for cash too. But as well as fishing and birds, I'm also into reptiles and there are plenty of those to see in Dorset – and then come August I am thinking of collecting and looking at mushrooms.

Which species of fish have you caught?

In the summer I like fishing for Tench and Crucians…

What?

Crucians.

I've never heard of them!

They are a species of carp, very small and ancient and quite rare now because they have the habit of interbreeding with Carp and Goldfish. They are a native species as far as we know.

And I fish for Barbel, which I started fishing for again this autumn and I used to fish for religiously, they were my species of choice. They are a very hard-fighting river fish, and in the past I fished the Kennet for them. I didn't fall out of love with Barbel but I fell out of love with the people who fish for Barbel because it all got a bit too serious.

They weren't pushing each other off cliffs or into rivers, were they?

You know, it all got a bit heated, yes. There would be exchanges and a lot of back-stabbing and it all got a bit too serious. It was too much to do with other fishermen and less to do with the fish.

The other side of it is that the River Kennet has gone into massive decline as an environment and that's quite hard to see – it's a bit like seeing a friend get ill. You don't like to see it and it's hard to watch.

And what's the cause of that?

Excess water abstraction is a big part of it. In the time I've fished the Kennet, the typical level has dropped by at least eight inches, and that's an awful lot.

But an even bigger impact has been the agricultural runoff – the chalk aquifers can no longer purify the water and so all that agricultural fertiliser goes into the river and there is a complete nutrient imbalance. There are algal blooms in chalk streams that ought to be a constant temperature almost all year round. The algal blooms knock out the plant life and the big beds of Water Crowfoot are gone, so you aren't getting the insects in the river and that impacts on the fish and the birds as well. The Kennet is certainly not a healthy river.

And the Hampshire Avon is similar: it has gone from being absolutely alive to a very sad state.

I can't say that when I worked for the RSPB I felt that fishermen, another rural wildlife interest group, were our closest allies. But as I listen to you talking about what has happened to rivers then the causes of declines in fish are at heart the same as the causes of declines in farmland birds – intensive agriculture. But there seems little common cause.

I agree, and you mentioned the Cormorant article I wrote. There wasn't any coming together of fishermen and the RSPB there. I think the law was just silly there because you could only get a licence to shoot 2–6 birds, and in some cases people were getting those licences just to shoot a few Cormorants almost for the sake of it, and in other cases, for example where you get a flock of eighty Cormorants eating Roach on the Hampshire Avon, where the birds would affect stocks, shooting 2–6 birds makes no difference at all. But the anglers wanted the Cormorant put on the General Licence and that shouldn't happen either.

The other thing I found frustrating was that it is much easier to blame something you can see, like a Cormorant, than all the things like extraction and nutrients that are much more difficult to see. In time the Roach will come back, but we aren't very good at giving things time.

Fishermen do seem quite keen on killing things, whether it be Cormorants or a growing clamour to cull Otters, which seems an unwise place to go.

Otters are probably more common than they've ever been…

Well, certainly commoner than they've been in our lifetimes – but they are doing a Buzzard, aren't they? They are coming back to places they used to be, and now some people are saying 'What a lot of Otters/Buzzards' – but they didn't spend forty years saying 'How few Otters/Buzzards'.

Yes, yes. The big difference with the Otters now is their impact on Carp. There are now gravel pits that are great places for Carp and these Carp can live up to sixty years and be sixty pounds in weight too – so they are magnificent fish. And these are people's livelihoods. I have a friend who lives on the Levels and he has a Carp lake with fish of over 50 lb which are irreplaceable, and there are Otters about and he has put an Otter-fence in. Now a dog Otter can take out a 30 lb Carp with no problem at all. And then, of course, the fisherman finds the remains with just the gill rake or a couple of vital organs eaten and the next day the Otter wants a fresh meal so it will take another one.

And anglers will say 'This is outrageous' – and people, generally, anthropomorphise far too much and they'll be describing Otters as vicious killers, whereas all it's doing is finding an easy food source.

The Eels have gone. They used to be the Otters' favourite food, but they've dropped by 95% and so suddenly the Otters are coming back to a very different world where there are no Eels in the river but they can go up a side-stream and find this dam with a lake full of Carp.

On a river, the river ought to be able to support Otters without a problem, but in some of these lakes and ponds I can see why the fishermen are concerned. And I think that Otters' habits have changed, because its more than thirty years since they were persecuted commonly and they are now much more diurnal and so people, including fishermen, see them a lot more. We see them in towns like Blandford Forum and Wimborne Minster on the

River Stour, and they aren't frightened of humans, and they have become an attraction.

It's very different if your livelihood is affected. So, for that friend of mine, if Otters take his fish then they'll be back again in thirty to forty years – but in the meantime he has lost his livelihood. It's very difficult.

Yes, I can see that.

There's one other point about Otters and that is that they might make a difference to Signal Crayfish numbers in some of these rivers where there are few fish these days. Where the Eel has gone Otters may lay in to Signal Crayfish and maybe make a difference to their numbers and maybe leave the fish a bit alone in future. That might work in anglers' favour – but maybe I am looking at it with rose-tinted spectacles.

I think we need to trust more to nature and let a balance form.

As well as being a fisherman you write about fishing. You've written two books.

The books I have written are *The Idle Angler*, which is more philosophical, and quite different from *The Twitch*. And the other book, *Rivers Run*, is more of a personal journey where the river is a bit of a metaphor for life – starting in shallow waters and ending up in the sea. It's much more reflective, and the aim is to appeal to a non-angler. It's about how fishing makes me feel. And I think the involvement with nature is one of the great things available to anglers which some don't choose to involve themselves in.

I've talked to birders who get excited about Kingfishers, but I have hardly ever taken my rod to a river without seeing a Kingfisher, and often really close. I haven't had one land on my rod yet, though other birds have, but one day one will. And I see lots of Water Voles and Grass Snakes.

I've said 'fisherman' a few times, and that's because my picture of an angler is mostly of a bloke – that's true often of birders too but my impression is that more couples go birdwatching together than fishing together.

Fishing is definitely male-dominated, but some of the very best fish have been caught by women. The record Salmon, 64 lb, was caught by Georgina Ballantine on the Tay in 1922, and the largest fly-caught Salmon by a woman called Tiny Morrison, which was 61 lb, also on the Tay and in 1924. There was a thought that women might be helped in some way by emitting pheromones, and that led to some men incorporating women's pubic hairs in their flies – it's not clear that it worked very well.

But I think that many women have the right aptitude for fishing. Although I don't want to generalise, some women are relaxed anglers and their nature is quite soft whereas some men are abrasive and harsh and they scare off the birds and scare off the fish too.

My wife, Sue, sometimes comes fishing with me. The last time she hooked a little Tench, about eight inches, very small, and I said 'Just take my rod and I'll sort that out for you' and she immediately hooked a 14 lb Carp which she played in perfectly…

And you were very pleased?

[*Through gritted teeth*] I was delighted. I was over the moon. And I was really happy that she had caught a fish bigger than I had caught at that lake.

What's your favourite fish?

It varies over time. It certainly was Barbel for years, and I've had two trips catching Barbel this autumn, and equalled my biggest-ever catch, so they have swum back into my thoughts. Perch I like – they don't grow particularly big but they look splendid with stripes and spiky dorsal fins. And I like Chub – they can be the easiest fish to catch and can be the most difficult. Izaak Walton called Chub the 'fearfullest of fishes' – he describes catching them with grasshoppers as bait and said the biggest fish in the shoal will always take the grasshopper if you present it over a shoal of Chub.

Favourite bird?

It's Buzzard. It meant the most to me when I was growing up as a kid in Hampshire because, apart from a few in the New Forest, we didn't see Buzzards. But when we headed west for holidays we would see Buzzards – they are the most incredible birds with that mewing call. Now I see them every day, but they are still my favourite bird.

But I like Robins too – they have a good attitude, and as a fisherman, wherever you go, if you have a pint of mealworms then a Robin will always find you and stay with you all day.

Your favourite book?

Lord of the Rings. I read it at an age when it made a big impression on me. And I suffered a lot with depression in my teens. When I was in a dark place, that book was a place to escape to.

Your favourite film?

Withnail and I – it's just brilliant.

Your favourite music?

Pretty diverse, including some classical and some heavy metal.

Your favourite TV programme?

Family Guy makes me laugh a lot. And I try not to miss Monday night with *University Challenge* and *Only Connect.*

TONY MARR

Tony Marr was born in the 1930s, was instrumental in putting Sussex ornithology on the map, and is now doing the same for the Isle of Lewis in the Outer Hebrides.

INTERVIEWED BY KEITH BETTON

I had no idea you were born in Scotland. Do you have any memories of your early years there?

Very few. I was born in 1939 in Glasgow, and I was only five when my parents left Glasgow for Sussex and I left my Scottish accent behind.

At what point did you become interested in birds?

It coincided with my going to Steyning Grammar School in West Sussex where I started as a boarder. When I was at school I had no more interest in sport than I do now, and I was allowed instead to go for country walks as long as I went with a classmate. I also became interested through my aunts, who lived in Skegness and drove me to nearby Gibraltar Point nature reserve in the school holidays.

Did the other boys at school react to your interest negatively, or were they accepting of the fact that you were different?

There were a few who saw that what I was doing was interesting, and we started a small bird club which we called the BPS (Bird Protection Society – obviously a rival to the RSPB!), but most of them used to take the mickey out of me. I was asked by the biology teacher to give talks to the other boys at lunchtime. I was given the use of the physics laboratory and a film strip projector, and I'd show pictures of birds and tell them all about them (or as much as the *Observer's Book of Birds* could tell me!). I think that helped to turn the tide a bit, but in those days it was a very unusual hobby. I was looked upon as a bit of a cissy to be going on country walks. I didn't have any binoculars so I just used to point out the birds in the trees.

Was there anyone else who was a major influence and focused your birding interest at that time?

I've always been grateful to three different men who found and fanned this spark of interest. The first was John Luker, my biology teacher at Steyning Grammar School. He encouraged this by allowing me to put up nest boxes and bird tables. Instead of looking at the blackboard inside the biology laboratory, I'd be looking at the Blackbird outside. He actively encouraged our ambitious school Bird Protection Society. We embraced the latest technology by investing in a rubber stamp which bore the silhouette of a Disney cartoon-style bird and our unique trinomial – 'BPS'. Later the society matured into the school's Ornithological Society.

My second inspiration was nearer home. My mother noticed a neighbour next-door-but-one to us in Southwick (near Brighton), a white-haired retired gentleman called Joseph Twort, who used to go off every morning on a bicycle with two bamboo canes strapped to the crossbar and a little bag slung over his shoulder (which turned out to contain his telescope and sandwiches). The canes were tied together in the middle and acted as a 'bipod' for the three-draw brass-and-glass 'scope. My mother learned that he was going birdwatching; told him that her son was becoming interested in birdwatching; and asked if he would mind if I went with him. He said emphatically that he would mind, as he wanted to be on his own! Mother, being very persuasive, finally talked him into taking me with him. It was a revelation.

From him I learned that there was a small local bird club called the Shoreham Ornithological Society, which started in January 1953. I became one of the younger members, and as a shy boy (yes, I really was!) I was delighted to find not only many motivating and inspiring older members, but that at the age of thirteen I was accepted among them. I soon found myself giving occasional talks, writing articles in their newsletters and annual reports, and I was even asked to lead small group outings. I also brought boys down from Steyning to some of the meetings. I learned so much from other people, and realised that here was an activity that was commendably purposeful, active, healthy, wide-ranging and great fun. Birdwatching was slowly becoming an acceptable pastime as we moved into the 1960s.

The Shoreham society was a model of its kind. It was run by Catherine and Jessie Biggs, two retired ladies who lived in Shoreham and were very public-spirited, bright and intelligent. They used their clearly formidable organisational skills and infinite charm to weld the whole entity together under a charismatic chairman, John Stafford, who was a local doctor. I became caught up in all this. It was exciting, it was breaking new ground.

My third mentor and inspiration was Harold Hems, a school-teacher from Sheffield I used to meet on my visits to Gibraltar Point. He spent his summer school holidays working as the bird observatory warden from late July to early September. He introduced me to the wonders of catching and ringing birds; identifying and photographing them around the reserve; and walks out on to the mudflats of the Wash. Harold opened up a whole new world to me. I realised that there was much more to birdwatching than I had ever imagined. These were very happy days, with a good regular crowd of local birdwatchers and others from further afield, attracted by the lovely surroundings, the great birds and the good company and birdwatching banter. Harold and his wife Margaret moved on his retirement to north Norfolk, and we were able to keep in touch regularly until his death in 2012 at the age of ninety.

Obviously other organisations did help to focus you, and you joined the Junior Bird Recorders Club in the early 1950s.

That was another inspiration, moving from a local organisation to a national one, particularly at their annual conferences, which were held at Belstead House near Ipswich. Other JBRC activities included bird ringing courses at Dungeness and Spurn; an annual essay competition; and bulletins and annual reports in which were published members' articles, papers and bird records. I made many new friends, several of whom are still active in national and international conservation in one way or another. For the annual conference we were all encouraged by the senior management of the RSPB to give talks to the other delegates, which was a great way of overcoming nerves and anxiety when public speaking. The quality of presentations was very high. Bert Axell, Dungeness RSPB warden, edited the bird records and produced an annual national bird report from the hundreds of records submitted by scores of members: a colossal undertaking. An annual prize (the Rainald Salzman Prize) was awarded to the member submitting the best work, consisting of a bird book of one's choice and a certificate. I was lucky enough to win this from 1954 to 1956.

So you went through your teens doing that, and remarkably in 1962 you were one of the founders of the Sussex Ornithological Society. How come there was no society in Sussex before that other than the Shoreham Ornithological Society?

This was the question that we all asked each other. 'We' included Richard Porter, Chris Mead, Bill Bourne (whose family came from Sussex) and Michael Shrubb. The answer was probably that bird recording in Sussex was dominated by Denzil Harber, who was the self-appointed county bird report editor, an operation he ran and financed single-handedly.

Sussex is blessed with a fine list of avifaunas over very many years, and we knew from them what the populations were of many of the species. We knew a lot about migration, but we had no idea of distribution for many species or whether certain birds were getting rarer or commoner. There was no forum for people to get together to share their passion and interest, other than the Shoreham society and a small natural history one in Horsham.

We were confident that the only way to achieve what we wanted was to get people together, offer them leadership and direction, and that volunteers would emerge.

As we had anticipated, people did emerge and could be offered posts in the embryonic society. We had a particular problem in persuading Denzil Harber to join us because he clearly enjoyed being sole county recorder and editor of the *Sussex Bird Report*, which he did very competently. However, his reports had one crucial weakness. He was very enthusiastic about watching migration and seeing rare birds and he loved nothing more than finding his own. However, the reports didn't publish any information about their numbers, distribution and breeding. Essentially the *Sussex Bird Report* was undemocratic, and a groundswell of dissatisfaction was emerging from a cadre of regular contributors who had decided that enough was enough. We did eventually persuade him to join us to become the society's recorder, but it took a great deal of patience and quiet diplomacy to achieve it.

The need for a county society was never in doubt, and was confirmed by the attendance of 250 people at the inaugural meeting in Brighton in February 1962.

I'm proud to say it's still going strong, and at the fiftieth anniversary conference in 2012 I gave an address on why, when and how the society was created and what we've achieved. I must admit that as the years slipped by, I did wonder from time to time if I would make it to the fiftieth anniversary.

Once we were satisfied that the society was founded on a solid base, we began a rolling programme to prioritise bird population surveys, gathering data and statistics on habitats and species which would enable us to tackle threats. These included proposals for yachting marinas in Pagham Harbour in West Sussex and Rye Harbour in East Sussex. These two harbours and their prolific wildlife were urgently in need of protection, but there was inadequate documentation of the wildlife. We were able to set up surveys very quickly, and after an immense amount of work by many people, including the Sussex Wildlife Trust working closely with us, both harbours were established as local nature reserves and have been running effectively and successfully for over fifty years now.

Denzil Harber wasn't always very accepting of other people's views. How accepting was he of yours?

Initially we got on very well, and there was a great deal of mutual respect based on our shared enthusiasm for bird identification and visible migration watching. 'DDH', as we all knew him, was usually very positive. For example, it was his idea to start migration watching at Selsey Bill, and Tony Sheldon, Mike Jennings and I, who by then were driving motor scooters, started regular and systematic watching there in 1959.

However, DDH was a controversial character who was loved by some and hated by others. I wrote an article for *British Birds* in 2003 providing a portrayal of the 'Sussex scene' in the 1950s and '60s and the legendary DDH's impact upon it.

Going back to the early days, what were your first pair of binoculars and your first telescope?

I was given my first pair of binoculars on permanent loan from a neighbour, thanks to my father. They were an old pair of wartime 8×32s. It was a revelation, to be able to see birds properly. And then my parents gave me a fifteenth-birthday present of a pair of Barr and Stroud 10×50 binoculars, which cost them £55, a lot of money then. They were huge and heavy but quite magnificent, and I used them for thirty-five years until they were stolen from my house when I was living by Grafham Water near Huntingdon.

It was a long time before I bought a telescope, and then it was one of the new prismatic ones mounted on a tripod which I had first seen in use at Falsterbo in Sweden in 1965 – nowadays a telescope is part of the birdwatcher's essential kit.

In 1970, at the age of 30, you joined the RSPB Council. How did you find the experience?

Daunting! Initially I felt overawed because at meetings I was rubbing shoulders with the great and the good in conservation, and at first I considered that I didn't have much to contribute. But I was nominated to Council as an ornithologist and an active birdwatcher, and I was in touch as SOS secretary with its (by then) 1,000+ members. I was about to retire from that post after nine years, and welcomed the opportunity for gaining an insight into national and international conservation. This I could see should broaden my vision, my view and my arena to the benefit of the RSPB too. I was there to a certain extent because of the work I'd done for the Sussex Ornithological Society and we had a good track record: Grahame des Forges and Michael Shrubb had both been on Council at different times.

I was most impressed by the quality of the RSPB staff, many of whom at that time were birdwatchers like me. This was perhaps the RSPB's greatest strength as it expanded: a core of very experienced, knowledgeable and committed staff who knew what they were doing.

The role of Council was essential, offering as it did a remarkably wide range of skills, talents and business and political experience, plus reflecting the views and wishes of the membership. The voluntary movement was growing fast, with members' groups springing up all over the country. One pivotal role was that of the Chairman of Council, whose job it was to act as a bridge between the directors and the Council members. We were lucky to have William Wilkinson and Stanley Cramp as chairmen during my first term of office.

Council members were appointed to a number of committees, and this was where most of the work was done before reporting up to Council when four times a year we met the management board. It was a rewarding experience, for me and I hope for the RSPB. I felt proud to be part of it, and was flattered when I was invited to return some years later and serve a second term.

The RSPB almost doubled in size in the time you were on Council – what was that like?

When I went back for my second five-year term on Council, in 1988 ('Centenary Year'), I was surprised at how things had changed in the intervening fourteen years. People were being brought in at a senior level with specialisms which seemed alien to loyal long-serving staff, but which clearly were much needed for a fast-growing charitable organisation – marketing, public affairs, parliamentary lobbying, media training, management training, facilities management, and other equally important aspects of running the business, which the RSPB had become. The sales department had grown enormously and was producing healthy profits. New directors had been recruited from outside the organisation, over the heads of long-serving managers who might have expected promotion. Membership was rising at a remarkable rate. Everybody was stretched to the limit, as was the building at The Lodge itself. Morale was suffering, particularly among some experienced middle managers who were reporting upwards to directors who appeared to know less than they did. The management board was becoming unfocused and was in need of stronger leadership at the top. These were challenging times.

My second spell on Council fortunately coincided with a period in my own life when my employer, the Land Registry, was itself growing and having to meet similar challenges.

Council was able to suggest principles and practices for the RSPB to adopt at a time when the management board was grappling with all these major issues. There were many serious political issues at the time, a dominating one being the proposed break-up of the Nature Conservancy Council, which took place in 1991. Direction and focus were restored when recruitment consultants were asked to identify a suitable candidate to head up the RSPB as Chief Executive, and Barbara Young was appointed in 1991. She brought management and motivational skills to the top table, restored the staff's confidence in the organisation and its future, and worked closely with Council and senior staff to lead the RSPB onward and upward. It was a very successful appointment, and I was very happy to have been a participant in such a crucial period of the organisation's development.

In 1992 you became involved with the BOU Records Committee – how did that come about?

I had a phone call one evening from the chairman, Alan Knox, saying they wanted a new member for the Records Committee and my name had been put forward. It was a complete bolt out of the blue. At that time, I'd become very interested in seabirds. I'd been on a small panel of so-called seabird specialists which Peter Harrison (author of the identification guide *Seabirds*, 1983) had set up to judge seabird records on behalf of the British Birds Rarities Committee (BBRC); apparently at that time few of their members had the relevant experience or knowledge, and I'd done a lot of pelagic voyages around the world. It led to one of the most interesting jobs I've ever done as an amateur ornithologist. As you well know, the British Ornithologists' Union Records Committee (BOURC) maintains the 'Official List' of British Birds, and adjudicates claims for additions to that list. The BBRC adjudicates all the other rare birds. It's a two-tier structure which I think works very well. Alan had made me an offer I couldn't refuse, and I was voted in.

At the time the BOURC had an image problem, and was often accused of being slow in reaching decisions and out of touch with the birding community. During my period as a member, and then when I became chairman for my last four years, we all worked hard to improve our performance and to try to be more accountable. We held the first-ever public meetings, open to birdwatchers and the birding press, well-attended and lively affairs; opened up the committee membership to a democratic system, inviting suitably qualified external applicants rather than hand-picked ones under the 'old boy' system; published regular and prompt press-release style notices reporting on our decisions; published annually a pocket-sized up-to-date 'BOU British List' for birders to use; and liaised more openly and more regularly with editors of the main birding journals. This shake-up was much needed and well received, although two or three of the longer-serving committee members were reluctant to emerge from the shadows into the full glare of such publicity. One of our most satisfying and successful appointments was that of Martin Collinson to the Taxonomic Subcommittee; he subsequently joined the main Records Committee and is now its very hands-on chairman, working closely with birders, committees and journals to speed up decisions (an increasing number, it seems) where specialised genetic analysis is required.

During your time on the BOURC did the committee ever reject a first record for Britain?

Yes, several. In fact, we rejected the first record circulated to me when I joined. It was a recirculation (or perhaps a re-recirculation) – hence the long period between the sighting and the decision. It was of a claimed Mottled Swift seen at Spurn Bird Observatory in October 1988. Opinions were divided between those who said it was a Mottled Swift (an African species) and others who suggested it was an unusually pale Common Swift. The Committee decided that there was insufficient evidence for this to be accepted as a first record for Britain.

Despite the scientific and serious nature of our work, we did sometimes see the funny side of things. At the top of the form which we each completed for every record, we entered our own experience of the species with a code. If I remember rightly, it was 'A = I have seen the species and know it well'; 'B = I have not seen the species for some time and am not very familiar with it'; and 'C = I have never seen the species'. Keith Vinicombe, a member of the Committee at the time, added 'D = I've never bloody heard of it'.

Looking forward, I suspect the role of the BOURC may be a little easier, because surely most of the birds now claimed as firsts for Britain are actually photographed?

That must be true, and the revolution in digital photography has had a huge impact on the evidence supplied to substantiate a record. Almost every birder carries a good-quality camera nowadays, and the technology continues to move ahead.

Have you actually been involved in finding a first for Britain?

Yes, the Sussex Greater Sand Plover at Pagham Harbour in December 1978. On a dull, damp December afternoon some birders came up to me and said, 'We've got a funny plover over here on the mudflats and we're not sure what it is – it looks like a gigantic Kentish Plover.' It was getting dark, and as I looked through my telescope I saw this obvious sand plover walking about on the mud.

I went back to Brighton where I was living at the time, got in touch with Alan Kitson and Richard Porter, and we met on site at first light the next morning. It took about two hours to satisfy ourselves that it was a Greater Sand Plover.

It was seen by over 1,000 observers up until 1 January 1979, when it probably succumbed to the prevailing cold weather. This was the first British record of Greater Sand Plover, and coincidentally the first British record of Lesser Sand Plover was also at Pagham Harbour, nearly twenty years later in August 1997, and identified by a local birder.

You got into seabirds in the 1980s, particularly off West Africa – is that what led to your becoming a tour leader specialising in the Antarctic?

It did eventually, but initially it stemmed from a private trip to Beidaihe in China in May 1993 organised by Colin Bradshaw, whom I was assisting. There I met John Brodie-Good, the managing director of WildWings, with Dick Filby, his then only tour guide. I had been on one of Dick's first Antarctic trips in December 1992 as a paying client, and loved the experience, particularly the seabirds on the Falkland Islands and on the crossings of the Drake Passage. In May 1994 I led WildWings' first tour to Beidaihe, which was followed by accompanying Dick as his deputy to Antarctica and finally leading my own WildWings trips there. At John's suggestion, I took early retirement from the Land Registry in 1996 and began to reconnoitre on behalf of WildWings a number of pelagic expeditions and land-based destinations which I then ran as tours for them.

Several trips to the Arctic culminated in two voyages to the North Pole on a Russian nuclear icebreaker and a final one into the Northwest Passage in Arctic Canada. Towards the end of my tour-leading career, which had included fifty trips to Antarctica and which ended in 2010, I began to prefer the Arctic, with its greater variety of species and habitats. My subsequent part-time move to the Outer Hebrides on retirement was heavily influenced by my wish to be as close as possible to my beloved Arctic and its wonderful birdlife.

You've led tours in many countries. Can you remember two or three incidents that stick in your mind as being good and bad (no names mentioned, obviously!)?

My most rewarding incident occurred during one of the six ship-repositioning voyages I made from Antarctica to the Cape Verde Islands, which called in at all the South Atlantic islands from South Georgia through to Ascension. Tristan da Cunha was always the highlight, with zillions of seabirds, three passerine endemics and the smallest flightless bird in the world, the Inaccessible Island Rail, to try to see. This involved landing on three islands, always fraught with problems as they are very exposed in the middle of the South Atlantic and subject to dreadful weather, which can make landing impossible. After outstanding seamanship and heroic diplomatic negotiations with the islanders, we made it, the first-ever expedition cruise ship to land passengers on aptly named Inaccessible Island and to see the rail.

My most unhappy experience was while visiting the famous Emperor Penguin colony at Snow Hill Island in the Weddell Sea in November 2009. I was the ornithologist on a Russian icebreaker for three visits, with 100 passengers on each trip who had each paid a very serious amount of money for this 'trip of a lifetime'. The first trip was delayed because we had a 48-hour blizzard on arrival, but we landed successfully at the colony after that. The second trip was delayed because the icebreaker became frozen in the ice (yes, an ice*breaker*!) for eight days – but then we got ashore to the colony for two perfect days. The third trip was by then delayed for more than a week, waiting for the ship to return to Ushuaia. We got back to Snow Hill as fast as we could, but encountered fog and could not fly the helicopters in to reach the colony, so the passengers never saw the hundreds of cute, adorable, photogenic baby Emperors which were to be the highlight of their expedition.

Finally, the funniest incident. A group of passengers on a landing on the Falkland Islands had arrived at a colony of Southern Rockhopper Penguins and were staring into the long tussock grass in which were hiding a number of well-grown young penguins. After several minutes of peering through binoculars and camera long lenses at these dark grey furry creatures, looking quite different from their smart black and white parents, they asked, 'Can you please tell us what these grey furry things are – do you get baboons here?'

You now split your life between Norfolk and the Isle of Lewis in the Outer Hebrides – why not Sussex?

Sussex is now too crowded with people and vehicles. After fifteen years of visiting many of the wildest and remotest places in the world, I need big skies and wide open spaces. Parts of Norfolk can offer that, but Lewis ticks all the boxes.

Is there a possibility that you might find a first for Britain again?

There has to be a good chance. The nearest so far is the first Chimney Swift for the Outer Hebrides (and third for Scotland), which was blown across the Atlantic to the Butt of Lewis in October 2014 by Hurricane Gonzalo.

If you do find a first for Britain on Lewis, what do you think it might be?

It could be something like a Willet or an American Oystercatcher, but even more than that I'd like one of the beautiful bright and colourful American wood warblers.

[Four months after Tony was interviewed, he found a Wilson's Warbler behind his house and over five days it was seen by 250

birders. Sadly, it was not the first for Britain, but it was the first for Scotland. So Tony's wish had almost come true.]

If you could have a whole day's birding somewhere in the world, where would you choose?

I'd choose Wrangel Island, in the far east of Arctic Russia, near the Bering Strait.

If you could meet someone from history, not necessarily a birder, that you could spend half an hour with, who would that be?

It would be E.C. Arnold, Headmaster of Eastbourne College, who wrote four books about birds which included *Birds of Eastbourne* (1936) and *Bird Reserves* (1940). He was an avid purchaser of land to protect wildlife, which included Salthouse Broad in north Norfolk, better known locally as Arnold's Marsh. He was ahead of his time in appreciating that bird protection was more a matter of buying land and protecting habitat than of making laws to protect birds. He was both a Sussex and a Norfolk man, and a great supporter of the National Trust. He had a wonderful sense of humour and hated bureaucrats and politicians. In every respect he was a man after my own heart.

Do you have a favourite film?

Goldfinger.

Favourite TV programme?

Have I Got News for You.

Favourite piece of music?

Beethoven's *Pastoral Symphony.*

Favourite bird book?

Wild America by Roger Peterson and James Fisher.

Favourite non-bird book?

South by Sir Ernest Shackleton.

Best place you've ever been to?

The Northwest Passage and Arctic Canada, for me the ultimate pristine wilderness. I saw it before oil and gas exploration will ruin it and the wildlife will be diminished by unscrupulous energy companies.

Worst place you've been to?

Barentsburg, the Russian mining outpost on Spitsbergen.

Favourite bird family?

Owls.

Best bird you've ever seen?

Emperor Penguin.

You've got an opportunity to see any one bird in the world that you haven't seen – what would that be?

Hyacinth Macaw.

TIM APPLETON

Tim Appleton is the co-founder of the British Birdwatching Fair (Birdfair), which has raised over £4 million for bird conservation. He was born in the 1940s.

INTERVIEWED BY KEITH BETTON

What are your earliest memories of birding?

I was born in Pembury in Kent but when I was very young we moved to Bristol, in the city but on the edge of a golf course. I lived there for twenty years. I was more interested in other animals in those days and spent a lot of my time with our pets – rats, mice, hamsters and tortoises. I was an outdoor kid. Often, I would find dead birds and bring them home. If I found a nest I might take the odd egg, and I really enjoyed breeding butterflies. I also used to rear badgers, so I was interested in wildlife rather than birds as such. I even appeared on the BBC TV show *Animal Magic* with Johnny Morris when I was fifteen, talking about animals in my area. I went to Wells Cathedral School in Somerset, which was a boarding school. My big thing was caving, becoming a leader at an early age taking adults down caves in the Mendips. That was primarily because being at a cathedral school I'd have had to go to church about four times each day. As long as I went to one service at 7.30 in the morning I could then disappear and I'd spend every weekend down caves. It was great fun. I was at Wells for five years and then moved to Bristol Technical College to take my A-Levels. I did not go to university because I was so wrapped up in everything else I was doing locally.

Did you have an adult who influenced your interest?

No, it was just me and not a case of going out with my parents. I was self-taught. My parents took me to Slimbridge in the 1950s and I developed a close affinity with wildfowl at an early age.

Did you keep a diary?

No. I knew all of the birds in my area but never made lists or kept a log of what I did.

Did you have a pair of binoculars?

Yes, I used some small Zeiss Jenoptems, and later upgraded to the 10×50 variety.

Did you join any clubs?

Because I was at boarding school I tended to just come home, see friends and go out in the woods.

What was your first job?

A lot of wildlife parks were being set up at that time, and in 1967 a man called David Chaffe set up Westbury Wildlife Park, so I volunteered to work there. I absolutely loved doing this because it was very much hands-on and it taught me a lot. Occasionally staff from the Wildfowl Trust would come to visit to sell us some ducks, and so I got to know Michael Garside, who was Peter Scott's personal secretary. One day I said to him that if ever there was a chance for me to work there I would like to be considered – it was my dream job.

At the age of eighteen I was busy running most of the activities in the wildlife park. A really good friend was Jonathan Silk, and he went to work for the Duke of Bedford at Woburn Abbey. An opportunity came up to work as a verderer, managing the deer herd. It was a phenomenal place with one head keeper and eleven under-keepers! I had an interesting time there, but it wasn't really the job for me.

In 1969 I received a phone call and was invited to be the assistant warden at Peakirk Waterfowl Gardens near Peterborough. I moved to live there, and for the first three months I mainly spent my time painting creosote onto the fence posts. I think they were testing me to see if I had staying power. But fairly soon after that I was rearing birds and really enjoying the job. It was a great place to be living, and at the weekend I would go out and catch waders with the Wash Wader Ringing Group. I was training to be a ringer with Tony Cook who was based at Borough Fen Duck Decoy. While based at Peakirk I got married and also became the warden. However, I finally got my chance to move to Slimbridge just after that and was appointed as deputy curator, working for Mike Lubbock.

It was an amazing time as I met so many fascinating visitors such as Jean Delacour and Roger Tory Peterson, and I got to work with people like Janet Kear and Geoffrey Matthews. In particular, it was great to work with Sir Peter Scott and his family. I specialised in hand-rearing of endangered species such as the Nene and White-headed Duck. In fact, I was the first person to rear Black-headed Duck in captivity. I was really in my element doing this – I was aged twenty-three.

What brought you to Rutland Water?

With the 1973 Water Act, there was pressure put on the water companies to take on the idea of wildlife conservation and recreation. Rutland Water was being created and the job of setting up the nature reserve was advertised by the Leicestershire and Rutland Wildlife Trust. Hundreds of people applied to win the job – as did I, but I was the lucky one and got it. The salary was £1,100 per year plus the use of a very nice little cottage, but I was twenty-eight so that was fine. I have now been there for over forty years. The reservoir was still very much in the early construction phase, and I worked with Dame Sylvia Crowe of the Forestry Commission to establish where the best areas were for tree planting and habitat creation. I remember sitting by Normanton Church on my first day in the job – 4 April. In front of me was a Spotted Redshank in its breeding plumage, and I realised how lucky I was to be doing this job.

There have been some good birds at Rutland Water during your tenure. Which ones do you particularly remember?

The best was a Bridled Tern in 1984. Thankfully quite a lot of people managed to see it but there was an incident which ruined the experience. One ardent twitcher, Franko Maroevic, trespassed to get a better view and flushed the bird. We never saw it again. He was arrested by the police for his actions but of course was let off. He did apologise to me many years later and we shook hands. Other great birds have been White-winged Black Tern in 1976, Collared Pratincole in 1977, Green-winged Teal in 1979, Ring-necked Duck in 1982 and Alpine Swift in 1993 and again in 2015. Of course, I love to see any rarities that visit the reservoir, but I don't twitch birds elsewhere, except maybe at Eyebrook Reservoir. I absolutely hate the idea of watching a half-dead bird with 500 other people!

You are seventy this year and have finally retired. What are you doing with your time now?

Yes, it's great. I plan to continue my involvement in Birdfair for as long as possible and I still work on it three days a week.

Let's talk about Birdfair – it's what most birders know you for. When it started in 1989, whose idea was it?

The idea began three years before. Peter Scott taught me that if you want to conserve wildlife you need to involve people. I started organising something I called a Wildfowl Bonanza. Martin Davies, the regional officer for the RSPB based in Lincoln, was doing a lot of great things such as running a Bird Bus, and he had a hide up in Sherwood Forest for people to watch woodpeckers. So we started to do things together. In 1988 the Game Fair was held at Belvoir Castle. I went along to see how it was run, and I thought about how it would be great to run a similar event for birdwatchers. I said to Martin, 'Why don't we take it another stage further?' We sat in the Finch's Arms in Hambledon and discussed it over a drink.

Neither of us had many business connections but we knew Bruce Hanson, who ran a major optical retail business, and approached him. He contacted a lot of the optical companies to suggest the idea of a bird fair and got a very positive response. John Brinkley at Swarovski leapt at the opportunity to be involved and gave us £2,000 – so that gave us the confidence to go forward.

We wanted three things: as well as a weekend where people could have fun, we wanted it to be a shop window for the industry, and we wanted to raise money for nature conservation. It was a success from the start and we managed to raise money quickly. In years one and two we raised £3,000 and £10,000, but we had no vehicle to take the money to manage projects..So, in year three we went to Birdlife International, raised £20,000, and the rest is history. It has just grown to become the world's biggest bird fair. It is the event that anyone involved the world of birdwatching needs to be at. We are also proud to see all the other bird fairs that it has spawned around the world. It is so great that the idea has blossomed so well in so many different cultures – all of which want to celebrate birds.

In the USA, it seems that bird festivals have been less commercial and more about social meeting. Is that true?

There is a booklet of festivals in the USA but they are often a gathering of 50–100 people. They have festivals in each state, and many different counties too – and some are simply dedicated to a particular species of bird. Getting a really big bird fair to work in the USA has been quite a challenge, partly because the place is so large and although there are many birders, they are spread widely. Bill Thompson, Wendy Clark and I talked it through and out of that came the first American Birding Expo in October 2015. They had about forty countries represented, it worked well, and it's now an annual event which changes venue each year. The problem in creating a bird fair is that organisers understandably want things to be a great success from year one, but you simply can't achieve things that quickly. Sadly, the Dutch bird fair has folded once again.

Looking at the British Birdwatching Fair, will it change much is the coming years, or is it now as big as it can realistically be?

It is almost as big as it can get in terms of space, simply because we are using every square metre that is available already! The people in the exhibition may change over time, and currently 58% of the stands are directly connected with travel and ecotourism. That aspect of the fair has grown, and that is where we see the most demand for new stands. The other limitation to growth is the fact that we have quite a small staff and their lives are taken over by this for six weeks of the year. We are so grateful to the volunteers who help put in a further 6,500 hours of time.

There must be a few things that have happened at Birdfair that you've managed to cover up and none of us knew about.

In 2012 we were really challenged by the weather. It was the wettest summer for a century, and other events such as the Game Fair at Belvoir Castle were cancelled just a month earlier and a lot of money was lost. Various rumours went around saying we were going to cancel it, but we made every effort to keep the paths and roads in good shape and everything went ahead exactly as planned.

Where's the best place you've ever been birding?

Without a doubt Argentina, and in particular the region of Patagonia.

Where's the worst place you've ever been birding?

There have been some pretty unpleasant experiences such as having an AK-47 stuck in my right ear by an Afar tribesman in remote southern Ethiopia in the late 1990s. I was looking through binoculars, which they obviously thought was a camera, and they were demanding money for taking photos of their cows. Having said that, I love Ethiopia.

I've always had a bit of an entrepreneurial side to me. I've always been on quite a small salary, and when I started at Rutland I used to hire 52-seater coaches, advertise locally and make 50 pence per person taking people to Minsmere, Welney and other reserves. I then did teaching at the local prison and did lifers' groups at

Gartree to earn money. When I was doing the local WEA courses I used to do day trips to Caerlaverock events and drive all the way to the Camargue. It grew and grew, so I had this group of people I used to organise holidays for (basically to places I wanted to go!) – so I travelled the world in the early days by funding it through this mini travel company.

Your best ever day's birding?

I'd love to say it was when I saw the Pink-headed Duck in Burma with Jonathan Eames and others. I've got no proof that we saw it, but it was one of my weirdest experiences ever – I'll never forget the moment when that bird peeled off from the Spot-billed Ducks that it was with, flew past us and went down into this massive reed bed. We spent days on elephants going backwards and forwards trying to flush it but we never saw it again. It was a moment of almost pure ornithological madness. Or crossing the frozen sea in Estonia in the worst winter they'd ever had and finding a flock of Steller's Eider.

Your worst day's birding?

Because I don't twitch I've never really had that experience.

Your favourite bird family?

Ducks, geese and swans.

If you had an opportunity to see any bird in the world, would you go for Pink-headed Duck?

No, because as far as I am concerned I've already seen it! Though I would like to see it on the deck. There are sixteen species of wildfowl I've yet to see, and because I've never really been properly into West Africa, one of the nicest would probably be Hartlaub's Duck.

Is there someone in the birding world you'd like to have met?

The person I'd like to meet again is Jean Delacour. He would have been about 85 when I met him – his four volumes of *Wildfowl of the World* were like the Bible for me. He was the nicest man, and very much the precursor of people like Peter Scott. I spent a day walking with him round Clères, which was at one time the world's most famous private zoo.

And somebody from the more distant past?

Probably Gilbert White, because I'd just love to walk with him in the English countryside, with the birds and creatures we no longer see or hear any more.

A bird book?

Bird Atlas 2007–11.

A non-bird book?

War and Peace.

Which film would you choose?

Ben Hur.

TV programme?

David Attenborough's *Life of Birds* series.

Piece of music?

House of the Rising Sun by The Animals.

TIM BIRKHEAD

*Professor Tim Birkhead FRS is a zoologist based at the
University of Sheffield who was born in the 1950s.*

INTERVIEWED BY MARK AVERY

Have you always been a birder?

Yes, but I'm not in the same league as other people you've
interviewed. And I have definitely never been a twitcher. I have no
idea how many species of bird I have seen in my life but I've watched
birds around the world, and studied them in many countries too.

My dad encouraged both myself and my brother to go birdwatching. We lived in a house on the outskirts of Leeds.

I had a friend at school, who is still my friend, and we used to
go birdwatching a lot. I went to Bramhope Primary School and
my teacher there, John Govett, was a keen birdwatcher. And one
day – he knew of my interest – he gave me one of those albums, of
cards from cigarette packets of birds, and I thought, 'Whoa! This is
just fantastic.' He also took me out birdwatching and he was good,
a keen member of Leeds Birdwatching Club which he introduced
me to. We had a project, when I was about five or six, when we all
had to read different bits of Arthur Ransome's *Great Northern?* and
we all had to illustrate the bits we had read. And I was given the
honour of painting the bird itself, and I was so pleased about that.
And that has remained an iconic book for me.

He took me to a local area, easily cyclable from where I lived,
Golden Acre Park on the outskirts of Leeds, and it was a pretty

wild sort of park. It was great for birdwatching, although I had no binoculars. I just cycled there and walked around looking at birds.

A memorable event at that time was that I saw a Great Grey Shrike kill a small passerine. I was kind of gobsmacked. The bird flew off and there was just this pile of brown feathers left so I got on my bike and cycled off to Mr Govett's house and told him what I'd just seen. He looked a bit askance and he came back with me and we looked at this pile of feathers and you could tell he was a bit miffed that he hadn't seen it.

Golden Acre Park was my patch for many years and I saw a Magpie roost there with up to 150 birds: it was very good for crayfish and I used to take them home to keep in aquaria...

Those were native crayfish? White-clawed?

Native. And there were some water birds such as Dabchicks.

I liked skulls and artefacts of birds and liked them as aesthetic objects as well as the fact that you learned something about biology from them. I remember poking around one time and finding this sack in a stream. And I opened it up and it was full of gamebirds and a lot of different waders. There were all sorts: Greenshank, Redshank, Golden Plover, Lapwing, Curlew and some ducks, and so I emptied this out on the bank and I cut all the heads off, took them home and buried them under a plant pot in the garden to dig them up three months later when they'd be clean skulls. It was quite a find.

Mum knew I'd done this and a while later she said, 'Have a look at this' – and there was an article in the local paper, and of course someone had found all the beheaded corpses and was wondering what on earth had been going on.

Your younger brother Mike was into birds. I knew Mike, at Oxford, before I knew you and he was studying Mute Swans then and is now a wildlife film maker. He was into birds in the same way?

He tagged along with me – I tolerated him! When we were at university our parents moved to Cley and that was just wonderful.

We spent a lot more time seeing our parents when they lived at Cley, and the three of us, Dad, Mike and I would often go bird-watching together. That was a wonderful time, and when after about ten years they moved closer to Norwich, I did miss those bonding experiences with my dad at Cley.

What were your first binoculars?

We never had that much money. Dad worked for a company that sold sugar and he was obsessed with the value of money. He said, 'If you want a pair of binoculars why not get a paper round?' And I did – in fact I did two or three paper rounds. My dad was very good at getting up in the morning and getting on with things, and I've adopted that stratagem too – when I wake up I get up and start doing something. I think the guy in the paper shop, who became a good friend, couldn't quite believe that I wanted to do three paper rounds before the other kids had started theirs. But it meant I accumulated quite a lot of money and I bought a pair of Ross 12×60s – they were enormous and I loved them. I don't have them any more, sadly, but I used them during my DPhil. I made a brass fitting for my tripod so that I could use them to look at Guillemots on the cliffs.

But then I flipped to Zeiss Jenoptem 8×30s which I still have, and I still like them. They weren't fantastic optically but they were very serviceable and very robust. Now I have Swarovskis and I love them. We use Swarovski scopes in the Guillemot study too.

How did you get into Guillemots?

Well, a series of serendipitous events, as is often the case. In my late teenage years, I became obsessed by Grey Herons and I used to go to a local lake about six miles from home, which was stuffed with fish and was a magnet to Grey Herons, and watch their behaviour – although they never seemed to do very much really, but they were very regal. And one summer I was working on my cousin's farm and he said he knew somebody who was studying herons and it was Ian Prestt. Although I didn't see much of Ian Prestt he provided me with this amazing opportunity to spend two weeks with his field assistant Tony Bell and go round lots of heron colonies. And when we finished fieldwork we'd go back to the office and sit and have coffee next to people like John Parslow, and as an eighteen-year-old I was chatting to all these people who were really interested and knowledgeable about birds. They invited me back and gave me the keys to the Land Rover and I spent two weeks studying herons and wrote a little paper on them for *British Birds*. I remember at the end of it saying to John Parslow that I'd had the most amazing time and if they were ever looking for someone to do fieldwork then I was their man!

Anyway, I went to university at Newcastle and, like many other people, I remember seeing a poster about a student conference at Oxford at the Edward Grey Institute and I went to it with my friend Rob Taylor. I gave this absolutely dreadful talk on herons and at the end David Lack said, 'If anyone is interested in doing a DPhil then come and talk to me' – so I thought I had nothing to lose and I went to have a chat with him. We walked up and down the road outside St Hugh's College and it was raining. I thought this was seriously weird but David Lack said, 'This is much better than being inside and I don't want it to be too formal.' He was very nice – but it was raining.

And there was another thread to this – at Newcastle we had had lectures about individual selection and sperm competition and we'd heard about Geoff Parker's pioneering work on dung flies and sperm competition and so I was just full of interest in individual selection and sex and sperm competition and David Lack just let me blather on in the rain, and he didn't say very much.

By the time I got back to Newcastle there was a letter offering me a DPhil place. I've still got the letter somewhere, and I couldn't believe my good fortune. And it said, 'We presume you will study Guillemots' – and later, a long time later, I discovered that John Parslow had had a word with Chris Perrins and David Lack and it had all been set up.

So at the end of my degree I spent a summer helping John Parslow with his study of Guillemots on Skomer. He didn't come but he set it up and left a load of instructions. He was interested in pesticides. What he'd done, bless his heart, was a mini outline for a doctorate, and it was just the most wonderful experience. I'd never knowingly seen a Guillemot before! But I just fell in love with Skomer.

Tell me about Guillemots. They are black and white seabirds, just like Razorbills, that sit around on cliffs – that's it, isn't it?

I've spent more time looking at Guillemots than I have any other bird and for me, this is a terrible thing to say as a scientist, but they are like people. A Guillemot colony is a town full of people gossiping, fornicating, squabbling.

And it's all in plain view – not like this street in Sheffield.

Exactly, and going into your hide and opening that window for me is like going to the cinema.

My DPhil was to try to understand the survival of Guillemots, because in the late 1970s there was a lot of concern about declining seabird populations and there was already someone studying Razorbills (Clare Lloyd) and someone studying Puffins (Ruth Ashcroft) and so there was a bit of a strategy there…

So Razorbills and Puffins were taken and you were left with Guillemots!

Well, I would initially have been delighted to study Puffins, and they are very beautiful, but they are the most boring birds in the world. They are so boring! And they live down muddy burrows. With Guillemots they nest in the open and you could do all sorts of observations. That was a monumental bonus to me. So I did the population study but I also studied their behaviour. In hindsight, the idea you could do a population study in three years on such a long-lived bird wasn't very sensible but I was hooked, and decided to turn it into a long-term study and I've been doing it for forty-five years and I don't regret it at all.

It's been great fun and it really did take about twenty-five years to understand the population dynamics: the survival, productivity, age of first breeding etc., and the population was at its lowest in the year I started. Then there were about 2,000 pairs to study and now there are about 25,000 pairs so we've seen a recovery, a big recovery, but in the 1930s there were probably about 100,000 pairs on Skomer.

You have almost convinced me that Guillemots are quite interesting, but everyone is interested in sex, and you have spent a lot of time studying sex in birds. There must be something interesting to say about that.

There is. I think I can tell you some interesting things about sex in birds.

We used to think that birds were all monogamous, but in a lecture I heard on sperm competition in dung flies...

Can you explain what sperm competition is?

Yep, so when females are inseminated by more than one male then there is competition between the sperm of the different ejaculates to fertilise the egg. And some of the original work on this was done by Geoff Parker on dung flies...

Those orangey flies that sit around on cow pats, and that's where they mate...

Yes. And Geoff had spent his PhD watching dung flies and noted that females routinely mated with several males. They key to Geoff's success was thinking about sperm competition in terms of evolution and natural selection operating on individuals. And in that lecture all the lights went on in my head and I really understood a lot more about evolution by natural selection than I had before. And the combination of a strong theoretical basis and sex was irresistible. I remember saying to my friend Rob Taylor, 'I'm going to do this in birds,' and he and several other people said, 'Don't be daft, birds are monogamous,' but luckily for me they aren't. And I started long before there were paternity tests available through DNA testing. So I first went through the literature and found lots of examples where people studying birds had seen females mating with multiple males. And these we now know as extra-pair copulations (EPCs). Some people had seen this behaviour and dismissed it – for example one of my heroes, Peter Marler, studying Chaffinches, had seen EPCs and had written it off as perhaps the male had a hormone imbalance.

As a mistake? The excuse of many a husband over the years, I guess. So do you think that most species of birds that we see in the garden and imagine as being paired up, are 'at it' at least occasionally?

Yes, certainly. There are actually a lot of common birds which haven't been studied in this respect, but species such as Great Tit and Blue Tit certainly go in for EPCs, at least a little.

When we started studying sperm competition we didn't know whether females could produce a clutch of eggs with mixed paternity – it seems odd now, but we didn't know. And because I had kept Zebra Finches in aviaries as a kid, I thought they would be good species to study in the laboratory. We used genetic plumage markers to study paternity. There are grey males and fawn males in Zebra Finches and if you mate a grey male to a fawn female you only get grey chicks, and if you mate a fawn male to a fawn female you only ever get fawn chicks. And with that system we showed that, lo and behold, you can have mixed paternity of a brood of chicks.

I then went through a bit of a crisis worrying whether what we had found might only be an artefact of captive conditions, so Keith Clarkson and I went out to Australia to study Zebra Finches in the wild. This wasn't straightforward because Zebra Finches tend to nest in remote places and they only breed when they want to, which is after the rains (and the rains can be unpredictable too), but I finally found a study area in northern Victoria where a guy was studying Zebra Finches. His name was Richard Zann and he sadly died in a forest fire years later but he was very kind to us and let us study his colour-ringed population of Zebra Finches. What we discovered was just like in the laboratory – it was just fantastic. It was exhilarating doing the fieldwork with goannas walking past and poisonous snakes running over your boots.

Another thing we realised was that you could use a bird's anatomy to tell you something about its mating system. So species where the males have big testes generally are likely to have promiscuous females. And there's a lot of information on that now. I've collected birds killed on the roads in spring for years, and there have been several PhD studies which have used those carcases, and that has helped us look at this issue.

I once went and gave a talk to a bunch of bird keepers in the upstairs room of a pub, and I can't remember exactly what I was talking about but it was going very badly. The vibes I was getting back from these guys were very bad: they were all sitting there with their arms crossed and leaning back in their chairs – it seemed like a disaster – I don't think they trusted this professor. So at half time I changed the talk completely and talked about testis size in birds and male sexual organs and a load of things like that and the atmosphere really changed, and they sat forward on their seats and they were very friendly. One of them came up to me afterwards and said that he bred Bullfinches and he had a bird in his freezer that I could have, so I collected it and when it was dissected it had the tiniest testes I've ever seen in my life (and I've dissected a lot of birds, and seen a lot of bird testes). This was very interesting so I talked to Ian Newton who had studied Bullfinches and he said that he had dissected lots of Bullfinches but they had all been immature males with small testes. In the end we discovered that Bullfinches really do have, for their body size, very small testes and we are pretty sure that this is because they stay in pairs most of the year and are not at all promiscuous.

But it isn't only testis size that gives you a clue – so do the external genitalia of some birds. If you see anything unusual in the anatomy it usually means there is something unusual about the mating system too. A friend of mine was studying Bearded Tits at Lake Neusiedl in Austria, and we caught these birds and I asked him whether he knew anything about the mating system and he said not really. So I looked at this bird and it had a funny cloaca. Now with some birds if you gently stroke the cloaca you can sometimes get a sperm sample so I did that and much to my surprise an enormous phallus flopped out and I said, 'These birds must be promiscuous' – and he went on to study them and sure enough they are.

Another example was that a friend of mine, Andrew Cockburn, visited Madagascar and saw a pair of Vasa Parrots and he wrote to me – I get quite a lot of letters which start like this, 'Knowing your interest in birds copulating...' – and he described seeing a pair of Vasa Parrots which copulated for a long time, so I contacted Chester Zoo who had some of the species and asked them. They said that male Vasa Parrots have a protrusion which goes inside the female and forms a lock a bit like mating dogs have, and the two can be locked together for forty minutes or so. And so I then had a PhD student who studied Vasa Parrots in Madagascar and they really do have an amazing system where every chick in the nest can be fathered by a different male.

And the last example is the Buffalo Weaver. I was reading about the Buffalo Weaver, a starling-sized bird in Africa that has a false penis in front of its cloaca – seriously weird! There is no other bird that has this. We found that that they have a bizarre mating system: a harem of about twelve females nest together in enormous twig nests and there are just two males for each nest. When we did DNA fingerprinting the two males sired most of the young at 'their' nests but there were also EPCs by males from adjacent nests.

So unusual reproductive anatomy turns out to be a good predictor of unusual and promiscuous mating systems in birds.

Not only are you a professor at Sheffield University but also you a Fellow of the Royal Society. The Royal Society was established by Charles II in 1660 and is regarded as the premier scientific academy in the world. So what does it feel like to be an FRS?

I was thrilled and filled with disbelief when I was elected to the Royal Society. I've basically done what I enjoyed doing all my career and I think I have just been incredibly lucky and it's wonderful to get this bit of recognition.

And that is what it is, isn't it, because you are elected by existing FRSs and only the very best scientists, across all fields of science, are likely to make it. So it's a bit like getting an Oscar for an actor. And it's always nice when the nice people get these things. How does it work?

Yes, you are proposed by two existing FRSs and then your name is up for election for seven years and if you get over a threshold number of votes you are in, and if not, that's it. I know a number of people who were keen to get an FRS and who missed out and it was a big disappointment for them, whereas I can say, with honesty, that if I had not been elected I wouldn't have been too disappointed because I had no expectation whatsoever of being elected. I still have to pinch myself these days sometimes.

How many fellow ornithologists are there in the Royal Society?

Chris Perrins, Ian Newton, John Krebs, Nick Davies and John Croxall.

Quite a bunch!

Yes, and in the University of Sheffield there aren't many FRSs. I think if you are in Oxford or Cambridge and you're not an FRS then you are made to feel guilty, whereas here in Sheffield I am made to feel special. It does mean that you get on a few more committees and things like that but also that your voice is heard just a little bit more.

Working at a university, which you've done for many years, is not all about sex in birds and jetting around the world to look at some more birds. You have to teach, and I always think it's a bit strange that these two jobs, scientific research and communicating science to young scientists, are rolled into the same job. You've seen lots of undergraduates over the years – have they changed at all?

I would say so. When I first started, undergraduates were much more diverse and more of them were prepared to take what I'd call an intellectual risk. They were up for challenge and were excited by ideas. Undergraduates are now tamer and more constrained and I blame the school system. I am appalled at the dumbing down that has happened. My view, and that of the department, is that when we get students to Sheffield University we have to get them to unlearn ten years of crap that they've had at school.

At school it is like they are buckets and they are being filled with stuff. When they come to university we want to light fires and inspire them.

I have always wanted to be an effective teacher. I was inspired by some of my teachers and I've always aspired to inspire students here. I'd like to think that I have been instrumental in changing the philosophy of teaching in our department and making it work.

When I arrived in Sheffield I taught all sorts of things including statistics and ecology, but now I teach two courses: animal behaviour to first-year undergraduates, and history and philosophy of science to third-years. And I try to make that third-year course a tough course: I don't let students take notes in lectures so that they have to listen and then go away and read about the subject, and we use innovative ways of assessing the students such as getting them to make videos or give a presentation. And the students, with hindsight, not necessarily at the time, really appreciate that approach, and I get letters years afterwards saying how much they can remember of that course.

But I need to say that if you work with the students and explain what we expect from them then they are incredibly receptive. At the end of three or four years some of our students are outstanding, and that is very rewarding for the teaching staff.

So, the way you talk, I am guessing that you think that other universities are not necessarily emptying the buckets and setting their students on fire in the way you do at Sheffield?

That's probably true. I just think that if you can teach your students to communicate – to write or to give a presentation – then that is such a transferable skill in life. It will help you through so many things.

And I guess communicating nature is what New Networks for Nature is all about?

Yes – I'm very proud of that. In fact, it is one of the things of which I am proudest. The idea wasn't just mine. A group of us, Mark Cocker, Jeremy Mynott and John Fanshawe started talking about setting up a celebration of nature which involved scientists, poets, musicians and visual artists, and we're now on our seventh year. It's been immensely stimulating and immensely rewarding.

Favourite places?

The high Arctic.

Favourite TV programme?

I actually hate television. The only programme I religiously watch is *University Challenge* – which will seem hopelessly nerdy.

Favourite music?

I like almost all types of music except choral music but I am a keen guitar player. I play the guitar for an hour every day. Rock guitar!

Really?

For my sixtieth birthday I treated myself to a Gibson Les Paul guitar, which is the Rolls Royce of guitars. Hendrix was one of my heroes, and Clapton, but the person I was most inspired by was Paul Kossoff of Free, who had a unique tight style and superb vibrato and I remember seeing them as a schoolboy in the Meanwood Blues Club outside Leeds.

Favourite film?

Far from the Madding Crowd – the first version with Julie Christie.

I saw that a few weeks ago on TV – it's brilliant, isn't it? I saw the new version recently too – that was pretty good and actually very similar to the previous version, even to the way the scenes are filmed.

Yes, I've seen the new version too. It's good, but I think the photography was better in the Christie version. I love Thomas Hardy as a writer; I remember being criticised at school for liking Hardy but I really do like his writing. I saw the first film several times – I can repeat much of the script word for word, and there's one line that relates to my research where Bathsheba and her maid are going to send the Valentine to farmer Boldwood, and the maid says to Bathsheba, ' 'Tis said he has no passionate parts.' I've spent a career studying passionate parts.

DAWN BALMER

Dawn Balmer is the Head of Surveys at the British Trust for Ornithology, and was national organiser of the Bird Atlas 2007–11. She was born in the 1970s.

INTERVIEWED BY KEITH BETTON

How long have you been interested in birds?

Pretty much all my life. I was born in Vennington, a small Shropshire hamlet, in 1970 and grew up two doors down from my grandparents, who were tenants of a farm, so more or less as soon as I could walk I'd be out with my grandmother taking lunch to my grandad in the field and looking at birds on the way. I first put pen to paper in 1976, though note-taking was intermittent in those days – it became more regular from the mid-1980s.

Was anyone else in the family interested in birds?

Not interested in birds in particular, but living on a farm my dad and grandparents knew all the common birds such Curlews and Yellowhammers. I have no idea why, but in the farm front room my grandparents had beautiful display cases with birds such as Herring Gull, Oystercatcher and other wading birds. I was always fascinated by these.

When did you get your first pair of binoculars?

When I was about seven or eight. I don't remember where they came from, maybe they were just in the house – they could have been my dad's. In 1988 I bought my first pair when I went to university and spent some of my summer job money. They were Vickers Adlerblick 8×42 and I think they were about £300, which was a lot of money at the time for a student. I upgraded to a pair of Leica 8×32 BAs when I started work at BTO and just last year changed to Swarovski 8×32 ELs.

Who were your early influences?

When I was twelve we moved from the tiny hamlet of Horsebridge to the small village of Pontesbury and I joined the local YOC which was run by the late June North. We had regular indoor meetings at her house, plus trips out locally and further afield, to see Red Kites in mid-Wales, and to Leighton Moss to see Bearded Tits. We were lucky kids to have a YOC group in our village. Also, when we were on family holidays I'd try to persuade Mum and Dad to go to nature reserves such as Bempton Cliffs and Kenfig Pools. A big influence on me was John Tucker. He worked for the Shropshire Wildlife Trust as a conservation officer and lived in the same village. He was very generous with his time and gave me opportunities to volunteer at the wildlife trust; we're still friends now. John got me onto doing my first bird surveys and atlasing in the mid to late 1980s. I also met the late Colin Wright when I was volunteering and we worked on data entry and compiling maps for the first Shropshire Breeding Bird Atlas together. Colin was so knowledgeable and inspiring. As we lived some way from Shrewsbury I didn't go to many Shropshire Ornithological Society meetings, so when I went to Polytechnic South West in Plymouth to do a degree in environmental science I joined their bird club and they – typically all male – were all very friendly and encouraging. There were lots of birders in Plymouth and an active club and social birding scene. Many of my student friends from Plymouth have gone on to work in conservation or birds, and remain good friends.

What bird books were you using at the time?

In terms of books, in the early days as kids we had lots of Ladybird books. I then had a first edition of Fitter and Richardson's *Pocket*

Guide to British Birds, which is still on the shelf now. Then I had the *Mitchell Beazley Guide* by Peter Hayman, and the *Shell Guide* which I bought in 1988 when I went to university and spent hours working out how I could see all the birds in the front of the book. I've got three copies of Lars Jonsson's guide – one in the house, one in the car, and one on the shelf at work.

Worst ever birding day?

On 21 October 1989 we went from Plymouth to Landguard for a Yellow-billed Cuckoo. We hired a car, got there in the late afternoon, but unfortunately two of our group saw the bird sitting in a dense bush and two didn't. I didn't. So we slept in the car overnight and of course there was no sign the next morning. I was really pleased to see the Yellow-billed Cuckoo at Porthgwarra, Cornwall on 23 October 2014 on the first day of our half-term holiday down there!

Do you still twitch?

Mainly round East Anglia and occasionally further afield, but with a young daughter it's not really a priority now (or possible!). Weekday twitching is completely out now, so I just have to hope good birds stay until the weekend. My husband and I did take her to see the Long-billed Murrelet when she was two months old – babies travel well in cars!

What's your British list?

472.

When did you reach 400 and what effect did it have on you?

399 was a White Stork in Thetford in April 1998. 400 turned out to be the Ramsey Island Indigo Bunting that I twitched on 20 October 1996 and added to my list in 1999 when it was accepted by BOURC. I'm sure I was quite pleased reaching 400, but my twitching has never really been about going for everything that turns up, because of work commitments and now I have a family. I'm sure I'll get to 500 though – still got a few 'easy-ish' species to catch up with.

During your time at university, did you manage to travel?

A bit. I met a few people in Plymouth who'd already left university, had cars and were living locally – people like Brian Field and Pete Aley – so managed to get around the southwest quite a lot. I started to do a bit of twitching so would go off with people from Plymouth and made my first trips to the Isles of Scilly in May 1989. I went to see the Golden-winged Warbler and Ancient Murrelet and hitch-hiked with a friend to see the Baillon's Crake in Sunderland. At that time, you could go anywhere in the country with National Express coach for £6 return, and I went with Mark Bailey from Plymouth on a 20-hour journey to Wick for the Harlequin Duck.

When you left university, what was the first thing you did?

Break my leg! After I left university in 1991 I took a summer job locally in Shropshire while I looked for employment, and in September I got run over by a lorry when I was on my bike – it was quite a bad break and set me back for about a year. I was very lucky to get a job with the BTO in November 1992 as a research assistant working on the Common Birds Census with John Marchant, analysing maps. That was a 16-month contract, so towards the end of the contract I started looking for other work and was offered a job with the RSPB as the Cirl Bunting project officer, which meant going back to Devon, which I loved. The BTO then offered me a permanent job so I decided to stay at BTO, where I really enjoyed working (and the Norfolk birding!). I stayed in the Census Unit for a bit, then moved to the Habitats Advisory Unit, which was a consultancy wing at the time. So I spent a couple of years doing lots of fieldwork in places such as Cardiff Bay, Barrow-in-Furness and North Wales, working a lot of weekends too. I missed quite a few rare birds during those years that I still haven't caught up with! The Bridled Tern at Foulney Island in June 1994 still hurts – I was working in Barrow but couldn't get there. There will be another I'll get to one day! Then I moved to work in the Demography team, running the Constant Effort Sites scheme and the Retrapping of Adults for Survival projects for a few years. In 2002 I started working on Migration Watch, which was really the BTO's first foray into the internet so it was hugely exciting to be part of. That led on to being BirdTrack Organiser in 2004 and various other things. In 2006 I went on maternity leave for seven months, during which time I applied for the Atlas Coordinator job, and I came back to do that in April 2007.

Having done the Atlas once, what would you do differently if you were organising it again?

Not too much actually, I think it was very successful. Bird Atlas 2007–11 was very much a team effort and we had a supportive steering group that provided advice throughout. Technology moved on quite quickly even during that time, and internet capability has developed so much further even since 2007, so I can imagine that data collation and feedback on progress during the atlas years would be even better. BTO announced that there would be a bird atlas starting in 2007 as early as 2004, which gave bird clubs the chance to think about local atlas projects. I thought our collaboration with bird clubs was good, and there is scope for improving that in the future. Although the atlas was successful in my opinion, it was also a massive task and very hard work, especially towards the end, so I don't think I'll be organising another one!

The statistical side of it was big. You drove people to do two-hour counts, and one of my observations was that people almost took that two hours to be a target rather than a minimum. Would there be a way of avoiding that next time round?

It's important not to confuse the requirements of a Britain and Ireland scale atlas with those of a local county atlas. For the Britain and Ireland atlas, the timed tetrad visits (TTVs) were a minimum of one hour or two hours. To be statistically robust we needed to repeat what was done in the 1988–91 Breeding Atlas, which first used TTVs. The idea of these visits was to provide information on the relative abundance of species; in a one- or two-hour visit you are most likely to record the species that are relatively common and more likely to miss the scarcer species. The aim of a TTV was not to produce a comprehensive species list for each tetrad. We used information from only the TTVs to produce all the abundance maps, so it is an important part of the design of the atlas. Tetrads were our sampling unit in a 10 km square, and we had an ambition to cover a minimum of eight tetrads in every 10 km square – which is a massive undertaking – so we had to have some minimum requirement to get the job done with good even coverage. Our aim was to have comprehensive species lists for every 10 km square, so the TTVs contributed to that but we also relied on roving records and information from other sources like BirdTrack. For local atlases such as the Hampshire Atlas, your aim was to compile comprehensive species lists for every tetrad, which required additional time and effort over and above the TTV. For me the most interesting maps in the atlas are those showing changes in relative abundance since the 1988–91 Breeding Atlas, and for them the data came entirely from those TTVs.

Maybe in future technology will make it easier – for example having on the screen a list for each square of the all the potential species and then people can see which haven't been found yet.

We made a conscious decision not to present species lists for 10 km squares from previous winter and breeding atlases at the outset. There have been so many gains and losses in terms of species, and we didn't want to influence volunteers in any way. Towards the end of the atlas, in order to direct volunteer effort to squares that needed more work, we did present information on the website to show a list of species that were recorded in surrounding squares but missing from the target square. I can imagine next time you could have an app so you'd know where you are from the location on your phone and if you're in a tetrad it would pop up and say for example, 'Mistle Thrush has not been recorded for this tetrad – have a look for one.'

The BTO has obviously changed during your twenty-three years there – what you have you noticed?

We've been making a conscious effort to better showcase our work and make information available to people, and the internet has helped enormously. I see us as being real leaders in citizen science, and using technology to enable us to engage with a lot more bird-watchers. A change in Director, from Jeremy Greenwood to Andy Clements, has also brought a change in direction and interest; we are keen to make the BTO better known and get our message out to more people. We are, and should be, very science-driven, but I think Andy's made us more aware that we need to raise our profile with a wider audience, and we've definitely seen that happen. We are now engaged with a much broader range of people and stake-holders. There have been big changes in sources of funding too, with the financial contribution from JNCC reducing significantly over the last fifteen years, so we've had to look for funding from a wider area. It does mean that periodic single-species surveys have been harder to find funding for and may become less frequent in the future. The BTO does have fantastic support, though, which has enabled us to raise funds for surveys like the House Martin surveys of 2015 and 2016 through appeals to members and supporters.

What would you like to see in the future?

I'm keen that BTO has a much closer relationship with bird clubs in the future; there are benefits for both BTO and the bird clubs. We hold several one-day conferences with bird clubs each year and these are always excellent events – and for me always an opportunity to meet some of our volunteers face-to-face. I would like the BTO to better support survey work at the county level, through advice on methodology, data collation and analytical support – we just need the funds to be able to do this. There are so many gaps in knowledge, particularly around population estimates of wintering and breeding birds. Local studies could be developed to provide information on densities across different habitats, which would be useful when we come to produce new population estimates at the British and UK scale.

There are a lot of female birdwatchers who are actively involved in birding but don't end up on records panels, etc. Is that their choice?

I suspect many of them don't get invited, or if they are, they might not have the time to commit to it. I'm struggling to think of many female county recorders in the UK – only Louise Bacon springs to mind. There is greater female representation on editorial boards of journals such as *Bird Study* and *Ibis*. I'm on the Rare Breeding Birds Panel as the BTO representative; my experience of recording and mapping rare and scarce birds for the atlas is a useful contribution to the panel. I also joined the BOURC in 2016 and have found that work interesting, but challenging at times.

When did you get into ringing?

When I joined the BTO in 1992, Jeff Baker was my first trainer, but I've ringed with many individuals and groups over the years. One of the best things I've done was to go ringing to Lesbos in 1994 with Jerry Wilson and his PhD student Antonios Kyrkos; we made six trips between 1994 and 1998 and caught a lot of interesting species. We collected detailed data on *Acrocephalus* warblers and Red-rumped Swallow and published two papers from the trips in *Ringing and Migration*. At the time, Lesbos wasn't really on the map at all and we had some great birding there too. When I came back I was speaking to Naturetrek at Birdfair and told them they should organise a trip there as it was an amazing place. They obviously didn't know me so didn't show any interest, but a few years later it became a really 'in place' to go. It's amusing that I later ended up leading trips for them to other destinations in Europe. I don't get a lot of time for ringing now, but really enjoy helping with the Nightjar project in Thetford Forest during the summer months. The use of GPS tags is giving us so much detailed information on the movements of Nightjars around the forest and some amazing local movements. My daughter, Bethany, has been coming out ringing with me recently so I hope to do a bit more in coming years.

The Atlas has been the biggest thing the BTO's done in the past ten years and you ran it. Did you ever feel it was unachievable?

It was a big task, but I'm immensely proud of being Atlas Coordinator. It was a dream job for me and a once-in-a-lifetime opportunity. It was a big team effort to collate 20 million records, and Simon Gillings was very much the brains behind the stats and the mapping, and we couldn't have done it without all the web technology, our Scottish and Irish organisers and Rob Fuller who kept us in order throughout. The volunteers were amazing as ever, some carrying out impressive amounts of fieldwork over the years and visiting far-flung places for the purposes of carrying out atlas fieldwork. Validation of the data was the biggest task and the one we underestimated the most. The last six months was very hard work, finishing the writing and editing all the texts. We set ourselves a target date for completing the writing, which I'm pleased we did, otherwise it could have dragged on. In recognition of the contribution the Atlas has made to nature conservation, the Atlas team were each awarded the RSPB Medal in 2014. That was quite unexpected!

You joined the editorial board of *British Birds* in 2003 – what is your role?

I'm currently secretary of the editorial board and I've enjoyed it from day one. Probably every other week or so we review papers that come around – not all papers get through and some have major revisions. I also keep an eye out for interesting local studies that would make good papers and encourage them to make a submission to *BB*. It's really rewarding when you see a local study carried out by amateurs (but highly skilled!) publish a paper on their findings.

Where is the best place you've been birding?

Oman – it's brilliant, so easy, the sounds and smells (cooking!) of the country, and the people are so friendly. The birding was just great, and best of all we hardly saw another birdwatcher.

Where is the worst place you've been birding?

The rubbish tip in Muscat, Oman, was incredibly smelly and had stomachs of animals piled up but was full of eagles and vultures. I've not really been anywhere that bad.

What was your best ever day's birding?

I'm torn. My only visit to Shetland in May 1994 to look for the Black-browed Albatross was memorable. It hadn't been seen for a few days but we went anyway to Hermaness and met some people who'd been there for the past few days and hadn't seen a sign of it. Incredibly, after a short time it flew in, but it wasn't just about the one bird, it was the whole place that made the day special. Another highly memorable day was at Killybegs in County Donegal in February 1999. I had gone with Pete to indulge in a weekend of gull watching and in one day we saw nine Glaucous Gulls, eight Iceland Gulls, a Kumlien's Gull, and found a first-winter American Herring Gull. Tolerating the fishy smell of Killybegs was worth it!

Is there a bird family that you particularly like?

Gulls, which I really got into in Plymouth. Nearby there was the Plym Estuary and a fantastic rubbish tip called Chelson Meadow which was a mecca for gulls. I was taught a lot about gull identification by Mashuq Ahmad and Dave Astins, who remain good friends of mine today. My second choice would be wheatears.

If you could go anywhere in the world, all expenses paid, where would you go.

Antarctica, not just for the birds but also for the landscape – I'd like to include South Georgia and the Falklands too.

If you could see one bird that you haven't seen before what would it be?

King Penguin.

You've met a lot of people in the world of birding, but there must be a few people you'd like to have met?

Peter Grant – his gulls book was very influential. Now I've lived in Norfolk for many years and have read quite a lot about the county, I would like to have met Richard Richardson on the East Bank of Cley.

Outside of birding, is there anyone in history you'd like to meet?

Explorers, people like Captain Robert Scott, Sir Ernest Shackleton and Sir Ranulph Fiennes.

What's your most wanted bird in Britain that you've not seen?

Gyrfalcon.

Imagine you're on a desert island and you could have a piece of music – what would you choose?

Move on Up by Curtis Mayfield.

What's your favourite film?

I was really into *Grease* as a teenager, but hardly watch films these days – there is just not enough time! The last film I saw at the cinema was *Dad's Army*, because it's my daughter's favourite TV programme.

Favourite TV programme?

Absolutely Fabulous – still makes me laugh.

A bird book to take to the desert island?

Greenshanks by Desmond and Maimie Nethersole-Thompson.

And a non-bird book?

Wild Swans: Three Daughters of China by Jung Chang. It is a family history that spans a century and tells the story of three generations of women.

JON HORNBUCKLE

Jon Hornbuckle has seen more species of bird than anyone else in the world. He was born in the 1940s.

INTERVIEWED BY KEITH BETTON

Where did life begin for you?

I was born in Bedford, but quickly moved to Nottingham and lived there until I left home for university. I had attended local primary schools but won a City scholarship at Nottingham High School via the eleven-plus exam. Much of my early youth was spent looking for birds and their nests, newts and butterflies on Bramcote Hills (sadly now covered in houses) and in Wollaton Park, near where we lived, on the outskirts of Nottingham.

Do you have any early memories of birds?

A particular memory was a magnificent Sparrowhawk knocking itself out by hitting the kitchen window before flying off after a few minutes of recovery.

Finding a few nests containing a Cuckoo egg was very satisfying, and I believed I found a Red-backed Shrike nest locally but I'm not so sure of that now.

My father was a science teacher and spent all his spare time painting or gardening. Unfortunately, it was many years before I developed a real interest in plants, and the move to the High School was no help for encouraging birdwatching. The school did have a train-spotting club, which I joined, and that led to me making weekly visits on my bike to Toton sheds and Derby Works, and then travelling further afield, such as to the meccas of Crewe and Doncaster.

So was this a bit of an obsession?

Definitely! Observation of new steam locos became an obsession with me for several years, and led to hitch-hiking and cycling all over England and Scotland to accumulate a big list of sightings. Eventually, the discovery of girls (and the demise of steam locos) put a stop to all this and I changed to the more sociable rowing, rugby, modern jazz and card-playing. I stayed on at school in the sixth form for an extra term to do Oxbridge exams and spent the rest of the school year at art college. Then I read metallurgy at St Catherine's College, Oxford. My main non-academic interest was films, and I eventually became president of the Oxford University Film Society. I tried to get a job in TV, the only way into the film industry at that time, then decided that a wife and baby meant money was a high priority so I took a job with United Steel Companies at Stocksbridge near Sheffield. I soon discovered the generosity of the steel industry, as Mr Peach, the chairman of the company, visited Stocksbridge to present all employees who had worked for more than fifty years – a surprising number – with a gold-coloured pencil!

What about your interest in birds?

My interest in birds was reawakened by a family camping holiday on the Yorkshire coast in 1970. I 'discovered' Bempton Cliffs and was fascinated by the breeding seabirds, especially the Gannets. I counted six nests, a far cry from the 10,000+ of today. I then set about trying to learn bird identification, initially by myself but soon with the help of Dave Herringshaw, a gifted Sheffield teacher and brilliant naturalist, ably assisted by the young Dave Gosney. They introduced me to the delights of moorland and woodland birding in the Peak District, and their enthusiasm for raptors proved highly contagious. I joined the Sorby Natural History Society and became a founder member of the Sheffield Bird Study Group in

December 1972. In the mid-1970s my job was based in Rotherham, so I sometimes visited the nearby countryside during my lunch break. On 3 January 1977 I noted a predominantly dark grey bird with a pale bill in a flock of House Sparrows of the same size. I presumed it to be an escape and did not realise it was a Dark-eyed Junco until a year or two later, whereupon I wrote to Mike Rogers, Secretary of the British Birds Rarities Committee, to ask if it could be a wild bird. Some months later I was surprised to receive a card congratulating me on having had the record accepted as the first ever wintering junco in the country (and still the only junco record in Yorkshire)!

Who was your main influence when you were a young birder?

There wasn't anyone, I was a self-starter. I was more of a naturalist, quite keen on looking for birds' eggs for a while and just taking the odd one, but I didn't make any real effort to look for 'special' birds. My passion for birds developed in the 1970s. It was probably when walking along the cliffs at Bempton that I suddenly realised how beautiful birds were, so I really became a serious birdwatcher at that point.

What were your first binoculars?

I didn't have any till I was a teenager, then I had the East German Zeiss Jenoptem.

First telescope?

I started out with the famous long Hertel and Reuss telescope. Since that I have used Nikon nearly all the time. I don't spend much money on binoculars or scopes, but I like to keep fairly up to date with cameras and lenses as long as they are not heavy or too expensive.

Were you involved in bird surveys at all?

Very much so. I threw myself into survey work, especially the Waterways Bird Survey which the Sheffield Bird Study Group (SBSG) pioneered in 1973 before it became a national BTO survey the following year. At first I surveyed the River Don near where I worked and then the Sheaf in Millhouses where I lived, but wanting a more interesting stretch, I tried the River Noe from Bamford to the start of Edale, in 1976. This I continued to monitor for the next twenty years, and I found it very rewarding. I also spent a not inconsiderable time on other surveys and studying a few of my favourite birds, such as Moorhens, Kingfishers, Great Grey Shrikes and Hen Harriers – the last two being more regular winter visitors then than today.

In 1977 I was invited to become secretary of the SBSG, a post I held till 1984. I served as chairman from 1985 to 1989 and joint annual report editor from 1983 to 1992. One day I was delighted to receive a phone call from Dr Tim Sharrock, then editor of *British Birds*, informing me we had won the 'Best Annual Bird Report' award for our 1991 report, a tribute to the efforts of many people, especially co-editor Simon Roddis. By then the group had embarked on a variety of surveys, including Sheffield parks, garden birds, Kestrels, rookeries and Starling roosts, and, most ambitiously, a successful tetrad breeding bird survey during the period 1975–80, all under the able leadership of Dave Herringshaw. This was largely possible due to the efforts of a team of keen youngsters, the like of which has never been seen since – Keith Clarkson, Ian Francis, Dave Gosney, Paul Leonard, David Marshall, Clive McKay and Simon Roddis being particularly active. The two Daves then embarked upon the considerable task of writing up the results, with the aid of a few like myself. They decided to broaden the publication to a complete avifauna of the Sheffield region, but after a couple of years ran out of steam. I then volunteered to take Dave Gosney's place and oversee the project to conclusion, little realising how much time this would take, given that this was before the home computer age so that the breeding bird maps, for example, had to be prepared by hand.

We finally went to press in 1985, by which time my main interest locally was in protection of Peregrine nests and monitoring of breeding Merlins. With help from the Derbyshire Ornithological Society, I organized the Alport Castles Peregrine nest protection scheme, a 24-hour watch which resulted in several years of successful breeding where there had been none successful before. On one memorable occasion, 11 May 1985, I stopped on my way to Alport to check the moorland near Burbage for Dotterel as I believed this to be a possible stopover site. To my delight I found a trip of twenty-seven, still the largest number ever recorded in the Sheffield area. By then Merlins were starting to reappear on the moors in spring, having become virtually extinct as breeders in 1960, due to pesticides. Along with a few other Merlin enthusiasts, I spent a lot of time on our moors in the late 1980s and early 1990s looking for and monitoring nests, under licence. I took up bird ringing with the Sorby Breck Ringing Group in 1988, another time-consuming activity, and so was able to ring a number of young Merlins from the expanding population. Several were subsequently found dead, near both the west coast and the east coast, with one travelling as far as Biarritz near the Spanish border – a great surprise to me although not unprecedented nationally. I continued ringing birds locally till 2013, when I reluctantly 'retired' for several reasons.

When did you become interested in international birding?

The seeds for this were sown back in 1971 when I had won a company award to study the steel industry in Japan. This became a three-month world tour starting in New York in February. The sight of a stunning Red Cardinal perched in a leafless tree in Central Park, followed by a hummingbird in Golden Gate Park, San Francisco, converted me to world birding. It was a poor time of year for birds in the parts of Japan I was able to visit, so I did not see much there, but my last port of call was India. I landed at Calcutta at night and was amazed to see many thousands of people asleep on the pavements as I drove into town. It was the time of the war between East and West Pakistan, which drove millions to flee across the border before the country of Bangladesh became a reality. A few hours later I was back at the airport for a flight to Assam to visit Kaziranga National Park to see the Indian Rhino. An afternoon drive on the only road into the park got me off to a good start with the sight of a hunting Tiger near a herd of deer! The following day I spent seven hours on elephant back, in two sessions, just a mahout and me, exploring the huge swampland with its birds and mammals including very approachable rhinos – not another tourist in sight, a far cry from today's situation. I knew I would return there one day.

Did you get to travel a lot in your career?

Apart from a job-related week in Nigeria in 1976, I did not travel beyond Europe again until my work changed in 1978 to involve technical liaison with customers abroad. My first foreign business trip was to India and Pakistan with sales manager Rupert Atkinson. He told me that his brother Rowan was just starting a career as a comic actor and had been booked to do a short piece on TV. Ours was a very eventful trip, including as it did a hijacking of our flight in Pakistan which could easily have ended in disaster – a man holding a hand-grenade pulled the pin out, detonating the grenade and blowing off his own arm! The pilot managed to return to the airport and we were able to continue our journey later in the day. The hijacker was executed a year later.

Did the rest of the trip to India work out OK?

Very much so. At a nature reserve outside Bombay, I was able to meet Salim Ali, the much-loved grand old man of Indian ornithology, who was opening a new nature trail. It was a great honour to have some of the birds identified for me by Salim Ali himself. The sight of so many new and colourful birds made me a tropical-birding addict for life. Not long after this trip, I went to Beijing on business and took the opportunity to arrange a meeting with Professor Cheng Tso Hsin, the Chinese equivalent of Salim Ali, another delightful ornithologist. It was not long after the end of the internal troubles caused by Mao; when I asked Professor Cheng how he had fared during this time, he said, 'Ten years completely wasted, I had to work in the fields like nearly everybody else.' By good luck Peter Scott also arrived in Beijing at this time and I was able to have a half-hour chat with him too – he had just returned from a long-awaited visit to the Russian breeding grounds of Bewick's Swans so was happy to tell me about that.

The new job enabled me to combine some overseas birding, mainly in the Orient and North America, with work. As my four children grew older, I started to use part of my meagre holiday allowance for overseas birding trips with a few friends, the first being to California and Arizona in 1985, then Costa Rica and two pioneering trips to Indonesia – Sulawesi in 1988 and Irian Jaya in 1991.

I had the opportunity to take early retirement with a modest pension in November 1993. The next day I flew to Ecuador for six weeks. I never did get the promised retirement dinner. I spent a lot of time in South America over the next six years, especially in Bolivia where I participated annually as one of the staff on a conservation research project, funded by Earthwatch and ably led by Robin Brace. I was responsible for catching and ringing birds, which was very rewarding as I caught 190 species including at least one new to Bolivia. I also undertook similar but shorter ventures in other countries, notably Peru, Ecuador, Papua New Guinea, India and the Philippines. I birded in sub-Saharan Africa fairly comprehensively and then concentrated on Southeast Asia and Australasia, with tour-leading in Papua New Guinea and the Philippines. With my love of travelling to new countries, helping in conservation projects and seeing new birds, I have made many friends and accumulated a big world list, although until the last couple of years it has not been my aim to see more species than anybody else.

Probably my most rewarding trip was to Abra Patricia in northern Peru in 1998. In the summer I did a major tour of northern Peru with friends and found that the old road at Abra Patricia had recently been resurfaced so that people were moving in to chop down some of the forest for the sale of wood and planting of crops. It looked as though this wonderful montane forest could be destroyed, like the lower-elevation forest had been, if nothing was done to protect it. I resolved to return to do a comprehensive survey on the birds and publicise what a great place it was but how it was in urgent need of protection. I did go back in November and with the help of Jeremy Flanagan, a British birder living in northern Peru, and a Peruvian birder for a few days, successfully did a three-week survey at four elevations and found many good birds. Living conditions were rather basic as we had to share our bedroom with some pigs, quite noisy at times, but we survived. We contacted a number of people such as Barry Walker and Robert Ridgely, all of whom offered support. Eventually, the Americans bought some of the land and built a new lodge, so that the forest seems to have been rescued and is now regarded as one of the top birding sites in South America.

Having achieved a major aim, to see more bird species than anyone else, and most of the iconic birds, what shall I do now? I've now passed 9,500. I don't want to spend much more of my limited funds in adding more birds to my list but I do want to continue visiting new countries. There are many eastern European countries still waiting for me, several 'Stans' in Central Asia, lots of Polynesian islands (although that's an expensive source), and a few interesting new countries in Africa, such as Chad and Somaliland, not to mention Iran and Egypt. So if my health stays good, I can see another ten years of action ahead.

In all your travels, have you ever felt your life was in danger?

I've been fairly lucky in that respect. I was on the hijacked plane in Pakistan but that was a business trip and nothing to do with birding. The only time on a birding trip when I didn't feel happy was the first time I went to Peru, in 1989. I visited Machu Picchu, and when I came down I walked along the railway line to go to a cheap place to stay that I'd heard about. Somebody followed me there and I did feel there was a possibility of being attacked so I didn't go by myself onto the railway line again! It was a bad time in Peru – nobody was going because of the Shining Path terrorists.

I was there by myself, and before I went I'd booked a ticket to Tingo Maria in central Peru. I'd heard that it was not safe so I got in touch with the travel-guide publisher Hilary Bradt – she was away at the time but a message came back saying it would be suicidal to go to Tingo Maria, whereupon I cancelled my flight there, and it was literally a couple of months later that a couple of English birders were killed there.

Do you put potential danger at the back of your mind, or do you just think you've been lucky?

If I have any serious worries, then I don't go, but I did believe that if I was careful and didn't really go to anywhere that could be a problem then I'd be safe. I didn't go to Peru lightly, but I hadn't been to South America and to me Peru was number one on the list. So I thought I'd do it, because I knew that Barry Walker was in Cuzco and I hoped I might be able to do something with him, which indeed I did. I spent the first week at the Explorer's Inn and met Ted Parker, I spent a week with Barry and a week with someone else, so apart from the day at Machu Picchu I wasn't by myself, as luck would have it.

I haven't got a lot of money but I have done a lot of travelling, so if I went for fairly high-quality accommodation then I wouldn't be able to do it. In the first years I was doing a lot by myself, sometimes with other people, and not employing guides because I didn't see the need for it at that stage. Now, I'm looking for the hard species – and for that reason I do tend to use good guides where I can. Trips are obviously costing quite a bit more now than they were in the past but I have to accept that because there's no point in going somewhere and missing the bird.

Have you faced any 'close calls' during your travels?

In China a couple of years ago I persuaded an American bird photographer living there to go with us as he could drive a car, and on one occasion he drove over a small bridge and very nearly tipped us all into the river, as he was on the very edge of the bridge.

Have you ever suffered badly with illness?

I've had malaria twice, both in Papua New Guinea, but not the deadly type. I've also had leishmaniasis, which I think I got in Bolivia – I had to have two weeks of treatment in my local hospital in Sheffield, and it took a while to clear up.

Over the time you've passed various milestones, the most recent one being 9,000. At what point did you reach the various stages of 1,000, 5,000 etc.?

I don't remember many of them, but when I reached 7,000 I was with a few friends in Indonesia and I wondered why one of them kept asking me how many birds I'd seen. Then when I identified what I thought was the 7,000th species Martin Kennewell fished in his bag and pulled out a bottle of champagne! I can't think of anyone else who would do such a generous act.

I used to report my annual world total list to the American Birding Association at the end of each year using Clements taxonomy:

Year	World List
2001	7,502
2002	no entry
2003	7,859
2004	8,073
2005	8,150
2006	8,230
2007	8,386
2008	8,401
2009	8,513
2010	8,693
2011	8,825
2016	9,595

By the middle of 2013 I was on 8,962 using the Clements list. I then made the decision to switch to the IOC list, like many others, and that added about 200 species to my total.

As you've gone through this journey from being a local birder in the UK to being a travelling birder, how have you noticed your thinking about birds changing?

I'm always pleased to see a new bird, some more than others. Most of my trips of late have been to places I've already visited, but not

Frank Gardner birdwatching in the French Pyrenees, aged 9 (1970)

Frank Gardner on a reporting assignment in Afghanistan (2003)

Frank Gardner on a BBC filming trip in Saudi Arabia (2013)
- photo by Dominic Hurst

Ann Cleeves (2015) – photo by Micha Theiner

Tim Cleeves with a Silvery Lutung at Kuala Selangor, Malaysia (2010)

Roy Dennis birdwatching (1970s)

Roy Dennis with satellite-tagged
Honey-buzzard (2007)

Roy Dennis with a Lynx in Norway
(2008)

Roy Dennis with a Beaver in
Bavaria, Germany (2008)

Kevin Parr on Mull (2005)

Kevin Parr fishing on the Purbeck
coast, Dorset, aged 8 (1970s)

Kevin Parr watching raptors, Sibillini Mountains, Italy (2007)

Tony Marr with Collared Dove
(1961)

Tony Marr birdwatching,
aged 11 or 12 (1952)

Tony Marr (2014)

Tony Marr (2015)

Tim Appleton in Spain (1995)

Tim Appleton with Bill Oddie
(1987)

Tim Appleton at Rutland Water (2015)

Tim Birkhead as 19-year-old
would-be rock guitarist (1969)

Above right: Tim Birkhead doing
fieldwork on Skomer, Pembrokeshire
(1974)

Tim Birkhead with Guillemot chicks
on Skomer, Pembrokeshire (2007)

Tim Birkhead doing
fieldwork in Zambia
(2009) - photo by
Claire Spottiswoode

Dawn Balmer (1975)

Dawn Balmer (2013)

Dawn Balmer twitching a Lesser Crested Tern on the Farne Islands with
Mashuq Ahmad and Chris Patrick (1990)

Jon Hornbuckle in Kakadu National Park, Australia (1995)

Jon Hornbuckle in Pisa, Italy (2007)

Jon Hornbuckle on horseback in Colombia (2010)

Tony Juniper in Malaysia (1991)

Tony Juniper at the UNEP Sustainable Finance meeting, Tokyo (2003)

Tony Juniper at La Brenne Wetlands, France (2009)

Tony Juniper ready for lobbying! St James's Park, London (2011)

Tony Juniper at Holme Fen, Cambridgeshire (2015)

Richard Porter (right) bird ringing at
Dungeness, Kent (1958)
– photo by Tony Marr

Left: Richard Porter aged 7 (1950)

Richard Porter birdwatching on
Blakeney Point, Norfolk (2003)
– photo by Steve Gantlett

Richard Porter in the Iraq Marshes
(2011) – photo by Mudhafar Salim

Richard Porter
with Iraqi
trainees, Iraq
Marshes (2011)

Bryan Bland playing Azdak in *The Caucasian Chalk Circle* (1950s)

Bryan Bland (2013) – photo by Penny Clarke

Bryan Bland going native in Norfolk (1970s)

Carol and Tim Inskipp in the Annapurna Conservation Area, Nepal (1986)

Above: Carol and Tim Inskipp near Machhapuchhre Mountain, Nepal (2012)

Carol and Tim Inskipp at Old Delhi railway station (2015)
– photo by Nikhil Devasar

Barbara Young as Chief Executive of RSPB, watching birds with Graham Wynne, RSPB Conservation Director (ca 1995)

Barbara Young as Chief Executive of RSPB, in fancy dress for Bird Fair (1991)

Barbara Young helping to cut reeds in Norfolk (ca 1997)

Barbara Young as Chief Executive of RSPB, visiting their Conwy Reserve (ca 1997)

Barbara Young (2015)

Bill Oddie (ca 1964)

Bill Oddie (ca 1964)

Bill Oddie (2009)

Bill Oddie with the third Goodies album (entirely written by him)
Nothing to Do With Us (1976)

necessarily the same sites. If I've been looking for, say, twenty new birds, obviously I'm keener to see some than others, but I do try to see them all. I don't go mad each time I see a new one, I just think 'I'm glad I got that one down.' There are some where I do smile a lot, but mostly the way I look at it is that I'm just searching to get the twenty or whatever. But importantly I only count birds I see – and of course some people count heard-only birds as well.

Who's number two now, and how will you feel if someone overtakes you?

It won't worry me at all. Either Phil Rostron or Hugh Buck. Phil will overtake me at some point as he's got the time and money, and I'm not sure he does anything else time-consuming, whereas I do spend a lot of time on birding but I'm going to cut down now to at least some extent.

What does it feel like to currently have seen more species of bird than anyone else ever?

I don't really think about it that much, it's just good to be able to go all round the world and see rare birds.

Do you have plans to write up about your travels?

I have actually made a start but haven't got very far, as it's hard work!

Do you know how many countries you've been to in total?

At least 120.

Is there anywhere you didn't like or was a disappointment from a birding point of view?

Not really, because I always do a fair bit of research when I'm going somewhere new.

West Papua must have been very tough in 1991.

Yes, but it's a piece of cake now! Our main problem was shortage of time, but we did walk up the Arfak Mountains in about one and a half days, which was quite hard, and there were times I was struggling to get uphill. I'm glad I did it, though, as it was a fairly pioneering trip. We did lose a day due to bad weather, which could have ruined the whole trip as we were on such a tight schedule.

Have you ever been on an organised trip?

The first one was in 2010 with Rockjumper on an exploratory trip to Angola, which I thought would be too difficult otherwise, and I finally visited the sub-Antarctic islands in 2013 on the annual Heritage Expeditions trip. I'm excluding at least ten trips where I was a leader for Naturetrek.

Have you ever totted up how much you've spent on travel?

I haven't. I managed to get into the Bolivia trip with Robin Brace, so every year for six years I went to Bolivia for 6–8 weeks nearly all expenses paid and then a few weeks in Ecuador afterwards, so a lot of my South America birding didn't cost me much. It's only in the past few years that I've spent quite a lot of money.

If you only had ten trips left, do you know which they'd be?

Not ten – there's only one ideal trip – Polynesia if I could afford it (which I can't). Otherwise I've mostly covered where I want to go, but I do hope to go to Iran and return to Brazil, Mexico, Borneo, China, Japan and Madagascar.

Favourite family of birds?

I love cranes and birds of paradise and am very keen on raptors.

Is there a family you're not keen on?

I'm not terribly keen on a lot of duck species, but there's nothing I definitely don't like.

Is there one bird you've had to have several goes at getting?

The one I was really keen to see was Long-whiskered Owlet. I put a few weeks into that in 1998 when no-one had seen it other than a few caught in mist nets on American collecting expeditions, but it wasn't till 2009 that anybody did see it in the wild, so after that I decided I must go back to see it, which I eventually did in 2014.

If you could see only one more new bird, what would it be?

I'll have two, please! Gould's Shortwing (probably in Yunnan) and Chestnut-shouldered Goshawk (in Papua New Guinea).

Do you have to budget for trips?

I normally avoid expensive trips, but having a certain amount of savings allows me to do what I feel like.

Any favourite music?

I do like early Bob Dylan numbers such as *Girl From the North Country*. I also like music by Crosby, Stills and Nash, Neil Young and Bruce Springsteen.

Favourite film?

Pat Garrett and Billy the Kid.

Favourite TV programme?

I really don't watch much TV but I do like the European series such as *The Killing* and *The Bridge*. Also the *Springwatch* programmes have been very good since Chris Packham joined the team.

If you could meet someone, dead or alive, who would that be?

I'd like to meet Sir David Attenborough – I've been to his lectures but have never talked to him personally. I'd also like to meet Michael Palin.

Best bird you've seen?

Steller's Sea Eagle. I went to quite a lot of trouble to see it in Hokkaido in Japan thirty years ago. I returned ten years later and saw about a hundred soaring over my head, which was quite an experience.

TONY JUNIPER

Tony Juniper is a writer and environmental campaigner. He was born in the 1960s.

INTERVIEWED BY MARK AVERY

Where did you grow up?

I grew up in Cowley, on the edge of Oxford. My father worked in the car factory there. But we were also close to the River Thames and Iffley Meadows, and within cycling distance of Wytham Woods and Port Meadow. As a not very old child I was cycling around there and looking at plants and butterflies and birds. It was a good place to be a young naturalist, in hindsight.

I went to school in Oxford but that didn't encourage my interest in nature at all – in fact it probably took away rather than added to it. Although, in my schooldays, there were lots of kindred spirits who did spend all their time cycling around and looking for wildlife – a great contrast to today. I think if you looked back at the lives of many of the leaders of the conservation movement today, they were also doing that as children.

Speaking to people who are close to the teaching side of conservation biology, I hear that many young people coming through into nature conservation today are people with a laptop and a knowledge of statistics rather than a deep knowledge of nature, which is very interesting. And I'm not sure we will see the implications of that for a bit in terms of how conservation works, but it's become a technocratic thing, to a large extent, rather than something that has brought people to it via a passion.

That initial experience does colour the whole perspective of nature, I think – it certainly does for me – and I wonder if we are approaching a time when future nature conservation leaders will come into their roles without that gut nature conservation feeling that comes from being a child.

Did you have any important mentors when you were young?

My parents encouraged my interest but they weren't necessarily particularly interested. I was taken fishing a lot by various people, relatives and family friends, and that was good. It's another aspect of the same interest for me.

The two things, fishing and birdwatching, definitely went together. The busy bird time tends to tail off in late June and then the fishing season starts.

I still go fishing, though I don't do as much as I would like. Every now and again I get a good day out: most recently on the River Test in Hampshire where in one swim which I have fished one day in each of five consecutive years I have now beaten my personal best catches of Grayling, Dace and Chub in one spot – which is a really exceptional bit of river. It's amazing!

And like many good fishing rivers it's full of birds: Little Grebes, Cetti's Warblers, Kingfishers – these things are there at the same time.

I'm always surprised that fishermen and birdy people aren't more cooperative.

Yes, they should be. There is an awful lot in common. But the two sides instead spend all their time arguing over Cormorants!

And often in the schism between fishing and nature conservation you see the animal welfare issue raised. That aspect made Friends of the Earth quite nervous – unreasonably so I think.

Fishermen certainly can be the strongest advocates for cleaning up water courses and reducing pollution. And there are a lot of them too! And anglers have a different demographic from environmentalists as a whole and so they are representing another group of views and giving a different political voice. But we've never really made that work, which is a bit daft.

Now the fishing interests have got their act together and they do a lot more campaigning themselves for water quality, which is really good because that will deliver nature conservation benefits for sure. It's really stupid that there isn't a stronger common lobby there.

I guess I was largely a self-trained naturalist really. I had lots of books – particularly the Observer series. I had the Observer's books of Birds, Freshwater Fish, Pond Life and others but my favourite, which still has pride of place on my bookshelf, was the *Observer's Book of Birds' Eggs*. That is by far the finest book ever published in the history of humankind.

After school I did some extra A-Levels because I decided I wanted to go to university – and I hadn't really thought about that before, so I needed some more qualifications. I decided that I wanted to do a mixture of zoology and psychology at a course at Bristol.

Robert Barton was a friend of mine, and a couple of years older, and he was probably the closest I had to a mentor in those days. He's still a friend. He inspired me to think about doing this course so I went off and did an A-Level in psychology and then did the course – which was a most enjoyable time. I have to say that I wasn't particularly devoted to academic studies as a young under-graduate. So I enjoyed Bristol, it was a very happy time, but not one that necessarily paved the way for a distinguished academic career.

At the end of the course I had two or three choices. Should I go off and do research? Should I go into the conservation world? Or should I do something completely different? And so I thought conservation was the best fit. This was the early 1980s – a completely different world from now.

The first thing I did, apart from a bit of volunteering, was to work for Berks, Bucks and Oxon Wildlife Trust (BBONT – Naturalists' Trust in the old days) working on education projects in south Oxfordshire where a handful of naturalists worked in schools. We were working with the same classes weekly over a year. It was really quite a powerful thing to do – and I'm not sure it would be possible now, with health and safety etc. – but taking a group of six-year-olds out, looking at birds, and coming back with twigs and plants and drawing them back in the classroom was amazing. Very soon when we went back each week all the kids would have a new story to tell about nature and what they had seen in the last week, perhaps in their garden. It was literally opening their eyes to the natural world and enabling them to see. That was a very positive rewarding thing to do – but it was only a one-year pilot and so that came to an end and I was looking again for things to do.

I did a few freelance projects and then did an MSc in conservation at University College, London, and then went to work for BirdLife International (then the International Council for Bird Preservation) on parrots.

Did you know anything about parrots?

Well, as a child, I had kept captive birds in an aviary. I had quite a diverse set of finches, Cut-throat Finches, Zebra Finches, and I had a few parrots too including lovebirds and cockatiels, so I had spent quite a lot of time watching parrots and, as a teenager, poring over books of parrots and I had quite a lot of enthusiasm for the group. So when I turned up for the interview I didn't know nothing!

The job was to write an action plan to save the world's parrots. We started with funding for a year but I ended up staying for over two years in Girton, to the north of Cambridge.

That was when Christoph Imboden was in charge?

Yes, Christoph was the Director at the time. When I arrived he made a big joke about whether I'd brought my budgies with me – but I'd brought enthusiasm for budgies if not the budgies themselves. So that was a happy period, putting together that plan.

And we did a pretty good job I think in raising awareness for parrot conservation. The definitive plan was published a few years later. Not many people knew that parrots were, pretty much, the most endangered bird group in the world. We did some campaigning and we did some research and that's when I went out to Brazil and we found the last Spix's Macaw in the wild. That was quite amazing.

And you wrote a book about it.

Yes I did, twelve years after all that work. That was a nice thing to do. But it was quite a thing to find the last individual of a species. The power of that story was that it told the decline, and the causes of the decline, of that species – and it was all made real by that one last parrot being found.

As with many parrots, it was a story of habitat loss and collecting for the captive trade, with a few other impacts of non-native species thrown in. But the reason for writing the book was to take that personal story and use it to convey a bigger message.

And all that time at Birdlife enabled me to get to know a lot more about tropical forests and where the very best tropical forests were, what was happening to them, what needed to be done – and then that was my way into becoming FoE tropical forests campaigner a few years later.

I spent eighteen years with FoE, and so tropical forests was my kick-off there with them. And that broadened into biodiversity generally and being the Campaigns Director and then the Director and the Vice-Chair of the international board of FoE.

What achievements at FoE gave you most pleasure?

I think the most pleasure came from putting the heat on people who should know better, whether they were ministers in government or international agencies or companies. Having the freedom and the knowledge and the resources to go out and say 'This is not right! You should be doing something different,' and being able to deliver that with some consequence, was always most rewarding.

Explain what you mean by 'with some consequence'.

Well – changes to the law, changes to supply chains, changes to how companies talk to their customers and the public. All these things came from pointing out that Company X is trading in timber illegally logged in Brazil or Government Y is knowingly allowing pollution to go into rivers.

And you got that by naming and shaming?

Naming and shaming and going out in public through an organisation that was beholden to nobody. We had total freedom to say what we thought was necessary to protect our biological system and that was a good thing to do. With small resources, looking back on that, the impact that we got out of those campaigns was really incredible.

To see the conversion of a bit of an idea into something changing in the world through putting together a strategy to change the world is quite a thing to do. It feels great when it works.

The two Acts of Parliament that we worked on, I look back on those as being quite important.

Those were the Countryside and Rights of Way Act and the Climate Change Act?

Yes. Those two legal changes stand testament to what you can do with not much money but a bit of imagination and a lot of enthusiasm.

Tell me about the Climate Change Act, because that was definitely seen by the rest of us as something that FoE led on, and that you led on. That's how all the rest of us saw it.

Well, thank you. I was pleased to be there and to have the opportunity to do the right thing at the right moment, but you never know that until afterwards. That period when I took over as Director in 2003, it was obvious that climate change was getting on the agenda.

And what you did was to get both of the main political parties on board.

Yes we did. And I was reminiscing over this with a friend, Sarah Mukherjee, who used to be the BBC environment correspondent, about the day when we got David Cameron into the media supporting a climate change act and we managed to blend the FoE brand with his for a while, which put pressure on the new government, led by Gordon Brown, to do more too. Brown wasn't that keen on the idea, not as keen as Blair, so we did party politics.

We managed to say, 'Here's the Leader of the Opposition, he's all for a climate change act, what are you going to do, Prime Minister?' And Brown said 'It's a very good idea and I thought of it first!'

And so we did all of that. And the person who really came up with the idea was Bryony Worthington, now Baroness Worthington. She said let's have an Act of Parliament that sets out what we have to do, taking account of the science. We said what a good idea, and we then had to reprioritise in FoE, which was quite difficult because we had to shut down several campaigns to free resources for this one, so it was a big deal. We called the campaign The Big Ask quite deliberately. And six months after we set up the campaign there was the setting up of Stop Climate Chaos, which was a whole gang of NGOs, and the RSPB was part of that (and your boss Graham Wynne was one of the co-founders), and that brought together a big coalition of environmental organisations, faith groups, trade unions and development groups into one big mass – and then The Big Ask and SCC came together – it wasn't always easy or harmonious but it happened and it all finished up very well because we got the Act of Parliament – the Climate Change Act. And the UK put itself in a position of global leadership.

And in recent years it feels like this has dropped away?

Oh yes! Big time!

How do you feel about things post the December 2015 Paris meeting? Because Paris was a big deal and a success.

Yes, well I went over to Paris with the Prince of Wales, who gave a speech. I travelled across with him and then stuck around for a few days and watched the closing piece. I do think that it is a very significant step forward. The world has been struggling with this for more than two decades and it's bloody hard. It's almost impossible to conceive of a deal which everyone can live with – it's not a simple subject, very far from it. And so the fact that we got the Paris agreement is a big and very positive jump. The question now is will countries do it and we don't know the answer to that but we have more chance now that we've got this deal than we had before.

We've all got a lever with our and other governments around the world. They've committed to doing something and now we have to make them. I think that's good.

I think Britain has let itself down in the period since the 2015 general election. And I don't think that this has much to do with money but more to do with the Conservative Party and the deep streak of scepticism over climate change that runs through their thinking. And as a result of scepticism about climate change the view that renewables aren't really necessary. And even if they were – they don't like them anyway! I think this is to do with a very vocal group of Conservative back-benchers who have managed to capture the Treasury and No. 10, and George Osborne seems very sympathetic to that view anyway, so we've seen energy policy written by the Treasury while Cameron is off trying to sort out Syria and the EU – he's not very interested in the environment, so we've got what we've got.

One of the sad things about Paris, I think, is that the public don't really believe that politicians are going to do what they've promised. There's a lot of 'they say that but they never do what they say' around, and I don't think that would have been the case, say, twenty years ago.

There is an element of that, and some of the green groups have said similar things – which probably doesn't help. It's usually best to talk things up rather than down.

But I am more hopeful. There is quite a lot that's changing and we have to take quite a long-term view I think. One thing that is very different is that industry is changing. We now have major companies on the energy side and consumers who are basically saying that this is the turning point. Big statements by 150 chief executives in Paris, standing on the platform and saying we are going to go zero-carbon. Even ten years ago they were saying 'Leave us alone. It's not true. Leave it to the market' – and that has all changed. The private-sector narrative has now flipped, and I think that is really important.

Many governments, particularly right-of-centre governments,

look to the private sector as their stakeholders – not the public. And now that the private sector is looking for a route to low carbon, governments will follow. And there is a generational change going on – increasingly tomorrow's politicians will have done climate change at school. And that will colour things too.

And so the cynicism that people feel is to some extent warranted, particularly with the likes of George Osborne, but things are changing.

What do you think of politicians as a whole?

Well, they are highly variable. There are those who are transparent and want to make things happen and there are those who are self-centred and ideological beyond being reasonable (whether left, right or centre). And we all know who the good ones are…

Well, I wasn't going to ask who you think are the worst, but name a few politicians whom you admire through your dealings with them over the years.

Caroline Lucas is an obvious choice. She is an old friend and now Green Party MP for Brighton Pavilion. She is an exemplary MP; hard-working, knowledgeable, principled.

Zac Goldsmith – I have a lot of time for him. Friends of the Earth used to collaborate with him before he was an MP. Of course he is from a different end of the spectrum from Caroline, being Conservative MP for Richmond Park. He has a lot of integrity and honesty. John Gummer – I always enjoyed working with him when he was Environment Secretary. I think the good thing about Gummer was that he started as a little bit sceptical about the environment, coming from Agriculture, but in a space of about six weeks he really got it. And became rather evangelical. He made a huge impact and generated a lot of respect. And in the process he did a great deal of good for his party. Modern Environment Secretaries from the Conservatives are just not in the same league. They just haven't understood how to build bridges with people who are not naturally Conservatives and generate a respect and standing with people who are not natural Conservatives, who might not naturally listen to them. Which is what Gummer did. He delivered a lot of benefits for the environment, for society and for his party and he didn't do any harm at all. If that's not good politics, then I don't know what is. Why they can't see this now I don't know. Liz Truss? I've not heard a word from her really apart from pork exports.

You haven't mentioned a Labour politician. You don't have to.

Yes, I'm just trying to think of one! David Miliband was very good in the period when we were trying to get climate change up the agenda and he was Environment Secretary before he went to the Foreign Office and he saw the read-across between the two briefs. Climate change and global security were two sides of the same thing for him. And he championed the Climate Change Act in the early days when we were trying to make the political case. He was very good.

You stood for Parliament in 2010 in Cambridge for the Green Party.

Yes, it was fun even though I didn't win. We tripled the vote and got the third-best Green showing in the country after Brighton Pavilion and Norwich. We went from 2.5% to about 8%.

There were moments in the campaign when it seemed like we might do even better – we had real momentum. The things that killed us were the first past the post system and tactical voting. A lot of voters said they didn't want the Tories, couldn't vote Labour, and that's why they voted Liberal Democrat, and got a Liberal Democrat MP, and then discovered they didn't like that either and went back to Labour in the last general election.

And you've lived in Cambridge for years...

Twenty-six years, yes.

...so you were hardly parachuted in.

No. I said to the Green Party that I'd give it a go but it would have to be Cambridge because I wouldn't feel comfortable anywhere else.

It was good and it was fun in hindsight, but hard work at the time. But I decided not to do it again in 2015 just because of how much time it takes up. I looked at all the things that I can do and decided it was too much of a commitment.

And one of the things you are doing is working with the Prince of Wales, isn't it?

I've been very pleased to help the Prince, and in my time as an environmentalist, one of the figures who was always speaking out on these things, and seemed unique compared with what the rest of the establishment was saying, was the Prince of Wales. He was giving speeches about whales, and rainforests and wildlife and everything else. So when I left FoE in 2008 and he said do you want to come and do a bit of work on rainforests I said I'd be pleased. So I gave a bit of advice to the Prince's rainforest project, and it's been a great success. It helped to reframe the discussion around that issue. It helped move rainforest conservation away from being just an ethical issue towards something that also made good economic sense because of water and carbon. He managed to pull off a meeting when we had the G20 leaders in London in early 2009 when they were going from Buckingham Palace to Downing Street he got them to pop in to St James's Palace for a cup of tea and sat them all round the table and got them to support a new rainforest initiative. And that in turn led to $4.5 billion being pledged after Copenhagen for rainforest conservation. I think you can legitimately say that his project did that.

The other stuff I'm doing now is quite diverse. It involves some writing, and I am involved with the Cambridge University Sustainability Leadership Institute, mostly with the private sector, and that's good, and, as of today, as we are sitting here, I am President of the Wildlife Trusts. And I need another day in the week to do that!

What do you think of the Wildlife Trusts?

I think they are doing extremely well in building their movement and doing stuff on the ground. One of the challenges for the Wildlife Trusts, and for all of us, is how we are going to get a stronger combined voice to get government to do more. We have a lot of combined support but quite latent, so it's a big challenge. Like any other big organisation there are a variety of views so it's quite a challenge.

I think that's one of the things we have to do in conservation more effectively over the coming years – to show politicians that lots of people do care and they want things done better, please.

I agree with you. Generally speaking, I think that's a role for NGOs, to mobilise their memberships to live more sustainably but also to act politically. It feels to me that almost every organisation is doing less of that now than ten years ago.

It feels like that to me too. I wonder what the cause of that is – because the need is higher. A couple of things I would reflect on is the extent to which many organisations set themselves massive targets for growth, and I wonder whether that has diverted them away from activism and into communications and marketing.

Another thing may be the technology. Everybody expresses themselves through Twitter, and that may have taken the edge off what the NGOs would traditionally have done…

…although I'd have thought it should open up great opportunities for NGOs too.

Yes, you'd think so. Another thing which I think has been important is the fragmentation of the media. A few years ago – you'll remember the same thing from your time at the RSPB – you would publish a report, go on the *Today* programme, and it would appear in quite a few newspapers and everyone would hear about it. That's much more difficult these days. It's very hard to get that sort of cut-through.

It might be that there are fewer Tony Junipers around who can get that media coverage?

Well, I don't know. But we are lacking some visible leaders in the conservation movement right now in that sort of activist space. We have some celebrity spokespeople, like Chris Packham, who's great, but they have a narrow space to move in because they have other jobs. It's difficult for them to slag off the Prime Minister without feeling the consequences

And sometimes they have lots of great passion, and the ability to say 'this is wrong' but not the technical knowledge to set out what should happen.

Exactly, yes, exactly. There is that area, isn't there, where you and I have worked a lot, between nature policies and politics and communications? And that is quite a difficult area to navigate – to understand how politics works and how to get things on the agenda and supported by the politics. There aren't many of us who have done it.

And even fewer who are doing it now?

Yes, yes. Maybe that competence has drifted down a bit.

Another thing that has happened a bit is that NGOs have got so exasperated with government that they have gone off and worked with business in business partnerships.

What do you think about that?

It can be good and it can be dangerous and a diversion.

I've never been that keen on it myself, because I've always seen it as so easy to lose any progress you make. With government, up to a point, if you get something you've got something for quite a long time, but business is much more subject to change. When senior members of the company change, then everything can change.

I think in the case of WWF the effectiveness of their strategy in policy terms has been hobbled by too close a link with fundraising. Many of the companies they have worked with have also given them money. Rather than choosing a sector (such as the food sector) in order to change the world they've tried a mixture of approaches. In practice, it's very hard – but you can work with companies like Unilever or Nestlé to deliver real change. That's different from going for corporate sponsorship and saying we'll give you a bit of advice and you can stick our logo on your stuff.

There's more to be done there. But I think some of what's been done hasn't been very powerful.

So I assume you aren't going to try to change the Wildlife Trusts into FoE? That would be a big step – but would you like the Wildlife Trusts to be a bit more outspoken?

Yes, I think I would, and I actually think they are outspoken. It's just that I wonder how much of what they say is noticed some of the time. But the Wildlife Trusts and FoE are quite similar in structure – a highly devolved network which it is quite challenging to get to act as a single national voice – so I have some familiarity with that.

Are you still a birder?

Yes, very much so. My parents spotted that I was going to be a birdwatcher, so I got my first pair of binoculars for Christmas when I was about seven or eight. They are Consort 8×30s which have lost a thread somewhere but they still work, more or less, and these travelled around with me for years. Then I had a pair of Carl Zeiss Jena which I had with me when I found the Spix's Macaw, and now I have a pair of Leicas and they work pretty well.

Do you have a favourite film?

The Dam Busters.

And a favourite book?

Endurance, the story of the Shackleton expedition to the Antarctic and what they had to do to save the crew after they got trapped in the Antarctic ice.

Favourite music?

I listen to almost anything really. The Rolling Stones are still a great favourite, but not much modern stuff.

Favourite TV?

I watch the *Ten O'Clock News* a lot, but I wouldn't necessarily call it a favourite, and I did enjoy the recent series of *The Bridge*.

RICHARD PORTER

Richard Porter is a conservationist who lives in Cley in Norfolk. He was born in the 1940s.

INTERVIEWED BY MARK AVERY

What is it like living in Cley?

It's great. I spend a lot of my time birding so I think I make the most of living here. My favourite walk is out to the end of Blakeney Point, which I do about three times a week. It's about a nine-mile round trip.

When did you last do it?

Today. The best birds were two Common Sandpipers and three Whimbrel. I was doing a Wetland Bird Survey (WeBS) monthly count – and I've been doing WeBS counts, under their various names, since the pilot ones in 1963, later to become the BTO Birds of Estuaries Enquiry, although we always just called that the Wader Counts.

How long have you been interested in birds?

Since the age of eight. I was given a book called *The Birds' Alphabet* by our next-door neighbour in Palmers Green, Miss Walcott. I've still got it, and looking at it now it was the most dreadful book with dreadful illustrations. Each bird had a poem, although for some letters there wasn't a bird and so for X it went 'X is for kisses to blow to all birds, to say we think they're too wonderful for words' – and there's an illustration of these little children all blowing kisses at imaginary birds. But the best one of all was Z, where it said 'Poor Z has no bird, Oh dear what can we do, I know let's take him for a day at the zoo' – and there's a little kid sitting on a camel holding a big letter Z.

No Zitting Cisticola then?

And no Zebra Finch either. The first one was 'A is for accentor, its eggs are sky blue, most people call it Hedge Sparrow, I do' and 'B is for Blackbird, its shape is the neatest, and of all bird songs, I think his is the sweetest.' But it was a book that really inspired me as a small child.

Where did you go birdwatching in those early days?

I used to go out with friends. I guess most of them aren't birdwatching any more. Occasionally my father would go out birdwatching with us. We'd go on the train to Bayford in Hertfordshire and sometimes just up the road in Palmers Green we'd go to Broomfield Park. And then I got permission to go into Trent Park from the Principal of Trent Park Training College and I had this official letter, and I think I was one of the few people allowed to go in, and every time I visited I had to write him a letter of what I'd seen and he'd always write back to thank me.

What birds did you see in Trent Park?

Well, I saw all three woodpeckers…

All three? I bet there aren't all three there now.

I bet there aren't. Breeding Red-backed Shrikes on the first visit I made, at the age of ten, this would have been in the early 1950s. And Spotted Flycatchers – I remember finding three nests, three active nests, in one bike shed. And in the rick yards, because this was a working farm in the college grounds, in the winter there were masses of finches and buntings – Bramblings, Yellowhammers, Corn Buntings, Chaffinches and Tree Sparrows.

From my home in north London I could walk to three pairs of Red-backed Shrikes in the early 1950s.

Wow!

I did my first bird note at the age of nine and a half, and I've still got it. And I've kept bird diaries all the way through from then until now.

What were your first pair of binoculars?

My father's 6×30 RAF binoculars with a webbing strap, and they seemed fine at the time. With money from my paper round I bought a pair of Barr and Stroud binoculars for about £46 in 1960.

What do you use now?

I've got Zeiss 8×30 and they are absolutely great.

Birdwatching may not be seen as the coolest thing to do these days as a child – in fact some people are teased for being interested in birds. Did you experience any of that?

No, no I didn't. But I think that was because, as a kid, I was quite well rounded and I joined in lots of things. So if there were any gangs I'd be involved, and if there were any scraps going on I'd probably be involved in them too, so I wasn't a loner and I wasn't seen as that kid who goes off birdwatching on his own – and that probably helped.

Did your parents encourage an interest in birds?

Yes, they encouraged me but they weren't interested in birds themselves.

The thing that is so, so different now is the freedom that we had. From the age of nine or ten I was going out birdwatching all day, on my bike, and going to some places that kids wouldn't be allowed these days because of health and safety. And from the age of about fourteen I was off hitch-hiking. My first visit to Cley was in 1957 on the back of the motorbike of Fred Lambert, who was a friend and a park keeper in London. He was about twenty-one and I was thirteen. And we stayed for a week in Salthouse and slept in a double bed – and that was how it was. Can you imagine that happening now? Back of a motorbike (I can't remember whether I had a crash helmet or not) and sleeping in a bed with a bloke eight years older than me? All sorts of questions would be asked about that.

My next visit to Cley was with Robin Jolliffe in 1959. We got the train to Hunstanton, he from Peterborough and me from London, and then cycled to Cley and camped in the field about 200 yards from this house.

I heard a story about that – that the two of you were in your tent overnight and heard something that might have been a Bittern and there was a disagreement about whether it was a Bittern or a cow. Can it possibly be true that a fight broke out over that?

That is absolutely true. In fact, Robin was reminding me about it a few months ago. We pitched our tent and there was this 'mooo!' noise and I think I said it was a Bittern and Robin said it was a cow – Robin was much more cautious than me in those days. So we argued. And the only way that we could resolve it in those days was to fight – and the tent came down and everything. But wouldn't that be a great way for the BBRC to settle disputes?

Who won the fight?

Well, I was quite a good fighter, but Robin went on to become a Cambridge Blue in athletics and rugby and had a game for England in rugby, so he was pretty powerful.

The version I heard was that he was quite surprised how well you fought back. Anyway, was school at all important in your interest in birds?

Not really. No it wasn't. I was very much self-motivated.

After school?

I went to the National College of Food Technology but I failed one of my first-year exams so I decided to leave. It wasn't until I was working for the RSPB that they let me do day release over four years and I did my degree, in Sussex, in ecology and animal behaviour. I doubt they would do that these days, what with the pressures on money, but I guess the RSPB thought that I was worth investing in. And it certainly helped me to get that qualification because it made me feel less 'naked' when representing the RSPB's views.

How big a gap was there between college and the RSPB?

Quite big. In 1966 I and three others went off to Turkey for the best part of a year and that was the start of everything, really, that has been important to me in my life. We did surveys of the wetlands in western Turkey and then we did the whole autumn, from July to November, at the Bosphorus – and that was the first ever comprehensive count of raptor migration ever done in the Old World.

Who had that idea?

I'd like to think it was me but it wasn't – it was my friend Mike Helps. Two years before we went he said 'Why don't we go off round the world birdwatching?' – but then changed his mind and said 'No, why don't we do something useful?' So with Ian Willis we talked about surveying the Dalmatian coast but we'd read a paper by Ian Nisbet and Chris Smout about raptor migration at the Bosphorus over a two-week period – and that was inspirational. So that's what we decided to do, count raptors at the Bosphorus, not go around the world, and we expanded it so that we spent the spring and summer in Turkey too. It took two years of planning and saving, and we also got a grant from the BOU. That's how it happened.

Alan Kitson joined us and we went off on the Orient Express with all our stuff including a holdall of film that Kodak had let us have cheap. We were met at Istanbul station by William Wilkinson who was working in Turkey with the Turkish Borax Mining Company.

And that must have been a large part of the impetus for *Flight Identification of European Raptors* – a Poyser book I bought when I was at school – I remember saving up for it.

Yes it was. There are many better books now but in its time, published in 1974, it was pretty good. I think its beauty was its simplicity and the great illustrations by Ian Willis. Although it was over forty years ago I got a royalty cheque for it a while ago for about three pounds! And it was translated into seven languages.

My job at the RSPB came through the Turkish connection. I enjoyed Turkey so much that I went there again with Alan Kitson for the following spring and then came back, worked on the docks, and then went back out again in the autumn. In September that year (1967) the IUCN, ICBP and IWRB had a big conference in Ankara on wetland conservation and I gave a talk about the wetlands we had surveyed in the spring. Now the Turkish government said that they wanted more surveys and Luc Hoffmann said he'd fund these and asked me whether I'd be involved. I said yes, but I hadn't really got a grounding in conservation, so I asked if it would be possible for me to go and volunteer at the RSPB for a while.

So that happened, except they paid me, and then the Turks didn't come up with their share of the money and so the RSPB asked me whether I'd like to work for them as a technical officer, so I said yes, and then the Turkish money did come through – and so for the next four or five years I spent half my time in Turkey and half working for the RSPB. I was blooming lucky. I didn't have a job interview or anything. If there is a case of being in the right place at the right time, then that was it.

And when was that?

That was 1968 at The Lodge. The RSPB had only moved its head-quarters from Ecclestone Square in Victoria, London, a few years earlier. I think there were about fifteen of us then – it wasn't many – it was certainly less than twenty. And we all sat around one table in what we called the Dining Room but is now the House Meeting Room and we all had lunch together.

I can remember lunch in that room when I started working at The Lodge almost twenty years later in 1986. There were more tables than just the one but we still ate together in that room and the food was wheeled down the corridor to us on a rather shaky trolley. Your office was upstairs and across the

corridor from where I shared an office, and I remember being somewhat in awe of you as you were one of the famous bird-watchers there at the time. I could hear your side of all your phone calls too.

I can remember some things being different in those days and some things being very much the same. Back in the 1970s I remember we had a very serious conversation about population growth – I'm pretty sure Gareth Thomas, James Cadbury, Nick Hammond and maybe John Andrews too were there, and we talked about human population and even wondered whether limiting family size should be part of the RSPB manifesto, as it were. It was a serious discussion. And we still haven't cracked that.

On a completely different level in those days too there were lots of Tree Sparrows nesting at The Lodge in nest boxes and we were worried about the fact that they were excluding Great Tits from nesting, so we actually modified the entrances to the nest boxes to prevent Tree Sparrows from using them all. Of course there are no Tree Sparrows at The Lodge now, and haven't been for decades. And on that subject I remember counting passerines on the tideline of the Wash during a winter wader count and we had a total of 1,000 Tree Sparrows feeding along the shoreline.

Funnily enough, we bought our house, here in Cley, from the daughter of Sir Malcolm and Lady Stewart who used to own The Lodge – so she used to live here and at The Lodge and I'm now here and used to work at The Lodge. When I talked to her son about this he asked me where my office was at The Lodge, and I told him, and he said 'Ah yes, the servants' quarters!' – which seemed fitting really.

What was the best bird you ever saw at The Lodge?

It might have been something as simple as Waxwings feeding in the cherry trees there.

And did you really have a competition with Gareth Thomas about who could reach the highest speed driving down the narrow drive to The Lodge building?

Yes we did! He had a really souped-up Cortina and I can't remember my car. We started the other side of the Sandy–Potton road to get some speed up and we needed someone trustworthy to say 'Nothing coming' – and then oooomph! Gareth won. I think he reached 105 mph and I chickened out at 95 mph, so he definitely won. And I remember him getting out of his car with a big grin on his face.

Peter Conder was the Director (I think he was called Secretary for a time) when I started work for the RSPB, and he had taken over after some goings on between Philip Brown…

Who went off to be editor of the *Shooting Times* – an unlikely career move for any future RSPB chief executive!

Yes!… and Gwyn Davies which were all very mysterious. I remember Irene Waterston saying 'If you only knew what had happened. If you only knew!'

So you worked for Peter Conder and then Ian Prestt, Barbara Young and Graham Wynne. What are your thoughts about the four of them?

Peter Conder was great. He gave you so much encouragement and support, and I can remember an occasion in my role as Technical Officer (I think it was Technical Officer (Species Protection), so I had all sorts of things like the Beached Birds Survey, investigating egg collectors and liaison with gamekeepers, and we all had to turn our hands to anything in those days) when there was a big oil spill (I think it was the *Amoco Cadiz* in Brittany) and I was helping with that and at the same time there was some report in the press about Chaffinches being deafened in some scientific studies at Cambridge to see whether song was inherited or learned and I had to talk to a journalist. He was trying hard to get me to say something about how terrible this was but I didn't, and instead I rather unguardedly said that we've got this big oil spill to deal with and I don't have time to talk to you any more about deaf Chaffinches – and this went all over the media and my name was mentioned in *Hansard* in debates in Parliament. I was expecting a rocket from Peter Conder but he said 'Don't worry, chum (he often called you chum), you carry on dealing with the press.' He didn't tell me off. He didn't need to, as he knew I knew that I had handled it badly. But that's something you didn't get later on – Peter Conder was special in that respect.

Peter Conder retired in his fifties because he said the RSPB was getting too big for him to run as a keen amateur and it needed a professional. Peter asked lots of us, including Gareth Thomas, Mike Everett and me, our opinions and we all said that Ian Prestt would be a good bet – and a few months later he was our boss. He was good and brought a formality to the organisation which hadn't been there before.

I guess as the organisation grew it went away from being a bunch of lads hidden away in the woods of Sandy to a bigger more professional outfit.

Yes, of course. That's right. And Ian did a very good job.

And then there was Barbara Young, who was brought in because of her management experience.

Yes, Ian wasn't well and was suffering from having had a serious accident. Barbara was brought in to take over and to put in management structures and make us all far more professional – and Ian knew that was needed. So Ian was moved upstairs (in the nicest possible way) to a presidential role. Barbara handled that very well indeed.

I think a great thing about Barbara is that she has continued to be a big supporter of birds and conservation after leaving the RSPB staff even though she knew little about it when she joined the RSPB. So she is a member of the Rare Bird Club and she comes to Birdfair and she still takes an interest in everyone and global conservation.

So how about Graham Wynne? Because you and I, at one stage, both reported to Graham.

Graham had considerable strengths, otherwise he wouldn't have been picked for that role. He was a good boss. I don't think he… I'm trying to think of the words… I don't think he got the best out of me as previous directors did. In fact, I saw him the other week along the Norfolk coast and he gave me a big kiss in the middle of Blakeney High Street. He was great! It was lovely to see him and what a great job he is doing now in the upper echelons of conservation thinking. The RSPB was a great training ground.

When did you leave the RSPB? I was trying to work it out as I drove over here.

It was in 1999. I retired from the RSPB early so I could pursue other interests, particularly Middle Eastern things. The RSPB kept me on a retainer and seconded me to BirdLife International for several years to help continue set up the BirdLife network in the Middle East.

What was that like? We hear about the Middle East a lot on the news, and it's rarely good news, and it certainly isn't about the environment. You've worked in all those countries, haven't you?

Yes, certainly a lot of them. They were really the halcyon days of travel in the Middle East. Yemen was pretty safe, but Iraq wasn't. Later on, places like Syria were wonderful, and Jordan always has been, as has Oman. And I'm still involved in the Birdlife Middle East programme as I know lots of the people and lots of the places – and lots of the history of it all. I've still got active conservation projects in Yemen and Iraq and BirdLife and OSME are shortly to publish the Arabic edition of *Birds of the Middle East*, which I'm really excited about.

The BirdLife model is that each country has a partner organisation – in the UK it is the RSPB – and that partner represents BirdLife and its work in that country. Now the RSPB is a big partner with lots of resources and so it gets involved in helping other smaller partners across the world. Do you think that model works?

I think it often does but sometimes doesn't. And the reason it sometimes doesn't work in the Middle East is that when an NGO gets big, the government shuts it down. Looking at the RSPB as your model for success just wouldn't work there, and I think, in the Middle East, the model has to be more of a partnership with governments to get things going.

Have you got a favourite Middle Eastern country?

I think it has to be Yemen. It was Turkey and funnily enough I was made a member of honour of BirdLife International last year and in my little acceptance speech I said that whilst my heart is now in Yemen my soul is still in Turkey. That's something I pinched from William Wilkinson, who became a great friend. I remember him, when he left the RSPB Council, saying that while his heart was now with the Nature Conservancy his soul was still with the RSPB – and I just pinched that because I thought it worked well.

You're on the board of *British Birds*? That's a great ornithological institution (as well as being a good read) as it is the journal of ornithological record.

I was on the board, I'm now a trustee. I think *BB* is wonderful and I enjoy being involved. I think there are two things that are really good at the moment. The first is *BB Eye*, which is a bit cutting edge and controversial. I'd like to see more of that because it's so easy to write bland things.

The other thing we've been doing in *BB* is a series on the UK Overseas Territories, and I think that, when complete, it will be something that it would be nice to pull together as a book, as it sets the store of what we've got and could help kick government conservation action.

They're a funny bunch of places, aren't they? Mostly little islands spread around the world.

But they are British! And we have responsibility for their conservation, which sometimes slips through the cracks of the system and I like reminding people of that.

Are you an optimist?

No, I'm not that optimistic. What we see around us we regard as the norm and we don't want to see it go any lower, whereas if we were alive fifty years ago we'd have a different experience. You've only got to go and spend a week in May in Poland and then come back here to realise what we haven't got. And yet most people you talk to, outside those who understand nature conservation, think the countryside around here is great. They don't realise that it is knackered. Educating people to understand that is terribly important, but I do worry that this whole business of climate change is taking people's eye off the ball of big issues such as pesticides and intensive farming.

How many bird species have you seen?

I'm very much a patch-work man when I go anywhere, but I guess my list is about 4,000, so not that many.

And which species that you haven't seen, if you could go anywhere in the world now, would you like to see?

Well, I think it would have to be one of the birds of paradise – and seeing them display has to be one of the greatest bird spectacles of all time. But if it were seeing a bird again that I have seen, then, speaking as someone who has found their own Slender-billed Curlew, I'd like to see one of those again – because it would mean that the species was still out there and not extinct. The one I found was on New Year's Day 1984 on the Yemen coast.

So I guess you weren't expecting to see one that day.

Not at all. It was at a sewage complex, and I crawled up to look over the pools and there was a group of waders in front of me and I saw this bird and I knew instantly what it was. No doubt at all, and then I spent the next five minutes checking that I wasn't being stupid.

Do you think they are still out there?

No, I don't, I'm afraid. That's one we've lost, I fear.

Do you have a favourite bird?

Yes, it would be Woodlark.

That's quite a classy bird.

And it's because of the nights I've had camping out on the Sussex heaths when you put your tent up and then this wonderful song starts. I think it is the best song of all and a super-smart little lark.

Favourite place to go birding?

I've always had a local patch wherever I've lived, and Blakeney Point is it now. And I've always been a bit paranoid about missing things there. It used to be the Lee Valley reservoirs and I'd regularly go down to Selsey Bill and I'd be really worried about what I might be missing there. I remember missing a small flock of Common Terns passing through the Lee Valley once and feeling really bad about it.

But in the world it would have to be Socotra. From a birder's point of view, the most important thing is that nobody else goes there – and since I've been going for so long and so many times, probably fifteen or more visits, I'm so excited about finding new things. It's like going to your own bird observatory, stuck out in the Indian Ocean, and knowing that few other people have the chance to go there. And I know that's selfish stuff, but I love my own company when I'm birding.

So with your local patch of Blakeney and your rather more distant patch of Socotra, what do you think about twitching?

Well, I have been a twitcher in the past and I guess I still do a bit of twitching so I don't have anything against it, but it's not really how I spend much of my time birding. I guess I was quite a keen twitcher decades ago, but you find that you have to go for everything or else you fall down the list – and as I fell down the list I couldn't really be bothered. I didn't want to be an 'almost' twitcher.

Recently I came across two old books by S. Vere Benson: *The Observer's Book of Birds* but also her *Birds of Lebanon and the Jordan Area*, which, importantly, was translated into Arabic – and we need to do a lot more of that, by the way. But she also founded something called the Bird Lovers' League – and the more I think about it, the more I think that's what I am nowadays. I'm a bird lover. With some of these things like shooting Hen Harriers or catching Blackcaps on lime-sticks in the Mediterranean I'm perfectly happy to say that I just don't like it. Even if killing didn't cause population declines I just don't like it. I put it in the same category as child abuse. I just don't like people killing or being cruel to birds. So yes, I'm a bird lover – and actually I feel we need to get more passion back into conservation.

Favourite non-bird book?

East of Eden by John Steinbeck – in fact all Steinbeck's novels.

Favourite music?

I'm pretty cosmopolitan but I love dance music – Gershwin, Cole Porter, that sort of thing.

Favourite film?

I think it's got to be *Casablanca* – have you ever met anyone who doesn't like *Casablanca*? *Casablanca* followed by *My Fair Lady* and another one, which might be a bit quirky, but it's so good a film, *The Exorcist*. Although *The Exorcist* is a horror film it's really well made and it is quite artistic.

Favourite TV programme?

Foyle's War (and before that *Dallas* and *Muffin the Mule*).

BRYAN BLAND

Bryan Bland is best known as a tour leader and artist. He was born in the late 1930s.

INTERVIEWED BY KEITH BETTON

At what age did you first become aware of birds?

When I was a toddler, my parents took me to Belle Vue Zoo in Manchester. As we entered the parrot house a gaudy macaw looked at me and said 'hello'. I was baffled by this and asked my mother how that parrot knew me – as I had never been to the zoo before and couldn't recall the parrot visiting our house. I must have been brought up not to speak to strangers.

As far back as I can remember I have been interested in all aspects of natural history – birds, butterflies, mammals, reptiles, flowers, trees and so on. I was brought up in Chesterfield in Derbyshire, so I was surrounded by the Peak District. When I was little I used to go off walking in the countryside alone. In those days parents didn't seem so over-protective. I caught and kept sticklebacks and newts, made fungal spore prints, collected skulls, practised taxidermy. I was keen on all wildlife but I was particularly fascinated by insects and especially beetles. I was frustrated that nobody else seemed to share my interest. I used to have to go to London to the Entomological Society if I wanted to talk to anyone about beetles. All of the wildlife notes in newspapers and magazines used to be about birds – nobody ever mentioned beetles. So by the time I reached eleven years old I had switched to ornithology as my main interest because there were other people who I could talk to about birds. Even more significant was the fact that studying beetles involved collecting them, and I couldn't kill a living creature.

What about your first bird book?

How to Study Birds by Stuart Smith, which was given to me by an auntie for my eighth birthday, so birds must have been an obsession even then. This is a book that is very heavy on text; not a child's book at all. It had a Blue Tit on the dust jacket but few illustrations other than histograms. At around that time a cousin gave me the *Observer's Book of Birds* and there have been a lot of Collins books added after that. In fact, in downsizing recently I parted with 2,000 books.

Who would you say was your main influence at the time?

Nobody in the family or my circle of friends was interested. TV had not yet discovered the appeal of nature documentaries – not that we owned a TV anyway – but there was Romany (the Reverend George Bramwell Evens), and Nomad the Naturalist (Norman Ellison) on the BBC's *Children's Hour*. And I still taste mint sauce when I hear the call of the Curlew (we were usually sitting down to lamb at Sunday lunch when that distinctive cry introduced my favourite nature programme on the BBC Home Service).

Did you join any club or society?

No, I did my own thing.

When you were at school were there others interested in wildlife?

After a gap of seventy years, I recently met someone who was a big buddy of mine at primary school. He said that his main memory of me was that I kept digging up dead bodies. There were other friends who joined me on walks but I don't think that they had a passion for nature. At grammar school I had a friend who would come out beetle hunting with me. He became an eminent surgeon.

What were your first binoculars?

Carl Zeiss Jena 8×30 – I can't remember where I got them from or when.

And first telescope?

I didn't get a scope until very late on. The first one was a Nickel Supra, about fifty years ago. Subsequently I have added a couple of Questars, a Leica and a Zeiss.

You were born just before World War 2. Do you have any memories of the war?

I remember being out in the fields doing the usual wildlife stuff and I thought a (British) plane was going to land on me but it came down elsewhere in the field. Nearby Sheffield was regularly bombed and we had an Anderson shelter at the bottom of the garden. When we had an auntie from York staying, and there was an air raid, she was dressed and down there in a matter of minutes, whereas we were all very casual about it.

Before you moved to Norfolk, where did you call home?

I lived in Chesterfield until I went to Manchester University in 1957 – when I was eighteen – to study English with history and French. After university I moved to London. I first lived in Putney, then with a family in Purley, then when I got married I bought a house in Oxted. My first job was with London Typographical Designers, and although I took a folio of my artwork down with me for the interview I quickly realised that the real opportunity was to help found a writing company. Although LTD were very well respected as an industrial advertising agency they had realised that, superb design notwithstanding, The Word comes first. So I left my artwork folio behind the settee in the boardroom and pitched for the writing job because they wanted to create a writing company – this eventually became National Editorial and Writing Services.

How many years did you run the writing company?

I became director of LTD (the core advertising agency) plus the associated writing company and also a 'thinking' consultancy. I was with them until I moved to Norfolk in the mid-1970s, but I resigned my directorships and became a consultant to the companies, the intention being to work alternate months in London to sponsor what I wanted to do, which was to set up a birding centre in Norfolk. It never occurred to me that the birding business would be profitable, but it took off immediately and within a year I resigned my consultancy.

You focused on an area 8 km around Oxted for five years or so – what was that like?

I didn't drive in those days and I decided to explore the area thoroughly. All my notes came to nothing in the end as during my final study year they built Bough Beech Reservoir within the 8 km circle, so the rarities I had seen suddenly became common. At nearby Godstone I found the first Red-necked Phalarope for Surrey in 1970, and also a Spotted Crake. On one occasion, when I found a Baird's Sandpiper, I realised I had forgotten my notebook so I had to submit my notes and sketches to the *BB* Rarities Committee on toilet paper, which I happened to have in the glove compartment.

At that time were you active in the London Natural History Society or Surrey Bird Club?

I was active in the 1968–72 Breeding Atlas and I was also a member of the Croydon Natural History and Scientific Society.

Before you went to university did you keep notes and sketches?

Sketching goes back as far as I can remember. My first published sketch was at age two – the art master of my brother (who was nine years older than me) wrote a book on child art, so my brother took along some of my work. I gather that at the age of two I should have been drawing a tree as a trunk with a big scribble on top whereas I drew an elm like the ace of clubs or cumulus cloud. When my parents died (within a month of each other) I found a cache of bird sketches I did when I reached the grand old age of three.

When did your hobby become your work?

I used to argue at management courses that the time would come when technology would take over and it would be a privilege to have a job and go to work. We should therefore develop training-for-leisure schemes, otherwise we would have a generation of malcontents who might turn to crime. Since I derived fulfilment from birdwatching I thought I'd set up a centre to teach enthusiasts how to get more out of this hobby. It coincided with the fact that my wife and I discovered that we could not have children. Although disappointed, we realised that one of the advantages was freedom of movement. Whenever I visited Norfolk I felt a sense of invigoration so we moved there. Cley-next-the-Sea was the obvious choice for a residential birding centre as it has always been Britain's birding mecca.

I see you found a Bonaparte's Gull at Durlston Head in Dorset. Can you tell me more about that?

That was in April 1970 with Jack Fearnside – a colleague from work, when we went for the Winspit Wallcreeper. The gull flew past and attracted our attention because it was unusual to see what we assumed was going to be a Black-headed Gull on the cliffs. A close view revealed it to be a Bonaparte's Gull (based on the underwing pattern, small bill and general daintiness).

Do you remember when you started going to the Scillies?

Over sixty years ago. I was camping in Cornwall with a school friend and suffering rain every day. Then we heard that there were these islands off Land's End bathed in perpetual sunshine. So we hitched to Penzance, camped on the cricket ground, got the old *Scillonian* over and stepped into a different world. We pitched at Sandy Banks (no Garrison campsite in those days), bent a penny under a logan stone, collected beads in the Beady Pool, spent a day re-enacting *Lord of the Flies* on Great Ganilly (William Golding's novel had just been published), swam naked across the lake in Piper's Hole on Tresco to mould Easter Island faces out of the putty-soft granite. We enjoyed it all so much that in September we sent a postcard from 'two Scilly asses' to the headmaster saying we were having such a good time that we'd be late back for the autumn term. It seemed quite reasonable at the time but when we did go home and back to school – by which time everyone had been back a week or more – I remember putting my head around the head's door to announce I was back. He replied, 'So I see, Bland, so I see.' He never mentioned the matter again. These days such behaviour could incur a fine. At university I persuaded half the English faculty – including my lecturer in American literature – to join me camping at Sandybanks. Then followed a lifetime of September/October visits, strictly for the birds.

Back in the 1970s we used to time our stays to coincide with the passing of the SS *Rotterdam*. In his early career the captain was saved by the Scilly islanders and as a mark of respect he said that every time he sailed past he would come close and salute them. He did this with every boat he captained until he became master of the flagship of the Dutch merchant fleet. Every fortnight it was going to and from America and he brought this huge ocean liner close in to the islands and tooted the horn. Naively we thought that all the American warblers hopped off his boat – so we'd look at the schedules and make sure we were there on a Thursday when it passed by. These days people sit at home, and when they get a pager message on a lifer they just go out for a day trip. It's not the same.

What keeps you returning there?

The beauty of the islands, the happy memories constantly evoked, and the fact that the social side of birding appeals to me as much as finding the birds – though discovering a rarity also gives great pleasure. I remember a visit in 1972 when, as I walked up the hill to pitch my tent on the Garrison campsite, I saw an Upland Sandpiper fly in from America. The next day with two friends I found an Aquatic Warbler. And on day three a Palm Dove (pended, following its treatment as an escape until a proven pattern of occurrence had emerged; two more were subsequently seen in Norfolk but it remains pended). Other finds range from Rose-breasted Grosbeak to Little Crake. One of the most recent (in 1998) was Britain's first Wilson's Snipe – though drumming up general interest was a slow and laborious task.

When did the work for Sunbird start?

Shortly after I moved to Norfolk. In the 1970s I had birded in Malaysia, Thailand, Mexico, India, Venezuela and Kenya, and also developed a three-week tour of Morocco in a long-wheelbase Land Rover with Mike McHugo of Hobo Travel. When Mark Beaman left Sunbird, and took with him some of the leaders, they asked me to help until they got sorted. Although I was fully occupied with my own residential courses (mid-weeks and weekends throughout the winter and full weeks in the summer) I agreed to take on some winter tours. When Serenissima Travel sold Sunbird in 1982 I became a director ('just until things are sorted'), then I took on more tours when Peter Grant died. Then I started the 'birds and music' tours that encroached into summer, and so life became more and more Sunbird, but still interspersed more or less back to back with Norfolk residential courses.

How do people react in foreign countries when they see you with a large beard?

When we were in Peru everyone was shouting 'Osama Bin Laden' at me, which I thought was because they were all non-hirsute and the only other beard they had seen was Osama on TV. But the following month in India the cries persisted, even though half of the Northern Alliance sported beards. I was quite lucky that nobody shot first and asked questions later. TV news bulletins reported sightings of Osama along the Pakistan border but I realised that it was probably me anyway.

I read that cuttings from your beard made their way into a Parrot Crossbill nest. Is that really true?

Oh yes. In 1984, when the first ever nesting attempt was being made in Norfolk, several of us were keeping watch on them and we soon realised that they were finding it difficult to locate decent nesting materials. So knowing that they like to use mossy hairs I cut off several chunks of beard and put them along the barbed wire. The birds were delighted and immediately collected the hair to line their nest. I don't think I can claim that their success was down to me, but it was good to help them out.

Have you ever had any disasters when leading tours?

For about a five-year period, everywhere I went in India I ended up in a different war. I went out with James Wolstencroft to set up a new trip in Assam and there was just a small window when you could get permits to go to these restricted places. Peter Grant was leading the Bharatpur leg then brought the group to me in Assam and we did Manas and Kaziranga. He was then bringing them back and I was transferring to do Sikkim but I never got the telex telling me not to go to Bagdogra Airport because there was a war on. The Gorkhas wanted independence. They'd blown up the hotel where we were due to stay and were shooting people in the street. I landed in Bagdogra and couldn't officially go anywhere else – my passport was stamped for two weeks in Darjeeling but that was the epicentre of the war. I did in fact go in a private car to what was to be our first reserve on the tour but of course I couldn't stop anywhere. So I was effectively an illegal immigrant till I got to the Bhutan border – but couldn't go there either, so I decided the best thing was to go back to the place I was 'valid' for, which was in the middle of the war. So I returned to Bagdogra airport where I was welcomed like an old friend, given an office. They asked me to teach some yoga, 'as you are so calm in this crisis'.

Our tour was rerouted to Kashmir and I flew to Srinagar. Truly Heaven on Earth. The following year Manas, just like Darjeeling, became a no-go area when the Bodos attacked, killed the rangers and took control. So a new Sunbird tour was arranged coupling Kashmir and Ladakh. Little did we know that, following publication of Salman Rushdie's *The Satanic Verses*, peaceful Srinagar would become the setting for more violence and, even more surprisingly, fighting broke out in Leh, the capital of Ladakh. I flew out in advance of the group to make sure it was safe but became trapped on the wrong side of the Zojila Pass when the road over the Greater Himalaya was blocked by thirty-five landslides. To meet the group in Delhi I had to run the 80 km over the highest road in the world (through sliding landslides). When I returned with the group to Kashmir and Ladakh we managed to avoid being stoned by the Buddhist (yes, Buddhist) aggressors – a saga in itself. And so it continued.

Once, in Egypt, we had a spare day at the end of the tour and I heard that there had been a Greater Painted-snipe at El Faiyum and decided to go there, not knowing that on that day in Cairo the CIA arrested an Egyptian responsible for the first bombing of the World Trade Centre. He was a follower of a blind ayatollah from El Faiyum so his supporters were looking for a reciprocal gesture. As we were driving out of Cairo a black limo drove beside us pointing guns, but our driver cleverly managed to get ahead of them somehow. Only the Egyptian co-leader and myself noticed what was going on – the clients didn't see the guns and merely commented on this very smart car which looked as if it was about to crash into us. At the El Faiyum turn-off we were given an armed escort. Eventually they stopped and said it was too dangerous to go any further because the next village was the home of the blind ayatollah. By a frustrating coincidence this was the stake-out for our target bird. Discussing the health and safety implications with our guards, I noticed that amongst the Little Stints in a roadside pool was a Temminck's Stint – a lifer for the Americans in the group. While I was focusing on that I noticed two Buff-bellied Pipits behind, which co-leader Sherif said were a first for Egypt. It seemed that this arbitrary spot was not such a non-birding location after all. So we walked out and (surprise surprise) flushed a Greater Painted-snipe from a reedy ditch. When we got back to Cairo we found out that a coach-load of Germans had been bombed. We were interviewed at the airport by the *Daily Telegraph* and the article appeared under the headline 'British birdwatchers unfazed'.

In Austria one lady in her eighties tripped going up some stairs in Forchtenstein Castle. I suspected that it could be a serious injury so I got permission to drive the minibus over the drawbridge, along the corridors and even up a flight of shallow steps to get near enough to lift her directly into the luggage compartment. The hospital confirmed she had broken her leg in two places and did the necessary repairs. The leading lady in that week's Eisenstadt opera took her presentation bouquet to her bedside and British Airways took out two seats so that she could fly back with us. But by and large I have been fortunate: it's all been pretty smooth.

Have you ever been imprisoned?

What a strange question. I once spent a night in Thurso jail because a kindly policeman offered it as an alternative to sleeping rough.

You have visited over seventy countries. Is there one other than Britain that you would rate as the very best?

India, Guatemala and Morocco rank pretty highly but it's usually the country that I am in at any given moment. Carpe diem. I would like to spend more time in Bhutan.

Is there a place that you have never been to but you would like to get to?

Papua New Guinea to see the birds of paradise.

Is there a bird family you particularly like?

I love Firecrests, Pallas's Warblers and all manakins, todies and hummingbirds – so I guess I consider that 'small is beautiful'. But then Resplendent Quetzals and Horned Guans are gasp-inducing too.

Is there a species that you would particularly like to see?

Any of those birds of paradise.

Has there been a day's birding (maybe in the UK) that surpassed any others?

There are too many to itemise. Maybe 30 May 1993 when north Norfolk offered Pacific Swift, Desert Warbler and Oriental

Pratincole as the star turns, with an incredible supporting cast. Nevertheless, the rarities meant nothing to a non-birder journalist who had joined my group to write an article on birders for *Country Life*. He was more interested in the Greenshanks. And why not? Rarities are, after all, usually common birds out of their comfort zone. Another rare threesome day included Red-necked Stint, Pallas's Grasshopper Warbler and Ortolan Bunting. In fact, Norfolk and Scilly have provided such a lifetime of superlative birding days that it is impossible to select just a few. Alternatively guiding the American James Vardaman in May 1982 in his quest to set a new British day list record was quite a busy day – involving driving my minibus around Norfolk in the morning, hiring a private jet to fly from Norwich to Inverness, a helicopter to Aviemore, a car around Speyside, and another helicopter to John O'Groats. It was strange to see both Bearded Tit and Crested Tit on the same day (plus Avocet, Scottish Crossbill, Spoonbill, Whooper Swan, Puffin and Black Guillemot). In retrospect the carbon footprint is embarrassing.

You have a big interest in music – where does that come from?

All the family (mother, father, brother, me) played the piano – and our next-door neighbour was a piano teacher – but we had no record player in the house. So I guess my love of classical music must be thanks to the good old Third Programme. I did have a friend at grammar school who owned a record player and I used to go to his house to listen to classical records. My tastes are quite catholic, but my devotion to Haydn has increased every year since first going to the Haydn Festival in Austria over thirty years ago.

Do you have a particular favourite piece of music?

Mozart's *Jupiter Symphony* or Haydn's *98th Symphony*. But if the scenario was as a castaway on an uninhabited desert island it would have to be something choral (say *The Messiah* or *The Magic Flute*, or even Richard Strauss's *Four Last Songs*) to remind me of the human voice.

Do you have a favourite bird book?

Birds Britannica by Mark Cocker and Richard Mabey.

Do you have a favourite non-bird book?

The *Compact Oxford English Dictionary*. I have the second edition – over 21,500 pages micrographically reproduced in one hefty volume. Hours of pleasure and endless food for thought. Etymology is as fascinating as entomology.

Do you have a favourite film?

No. For most of my life I have avoided cinema. I have always thought that a live performance (the theatre or a concert) is preferable – something that is happening at a moment in time and can't be repeated. But a recent visit to a modern cinema revealed to me that watching a film on the big screen is a totally different experience from seeing the same film on TV or on a plane.

Do you have a favourite TV programme?

Of course all of the David Attenborough documentaries. But I also like *Only Connect* and *Would I Lie to You?*, and most British dramas (*Inspector Morse*, *Lewis*, *Judge John Deed*, *Silent Witness* and *Spooks* are extraordinarily well crafted and eminently watchable).

If you could meet someone in the natural history world you have never met, who would that be?

Sir Thomas Browne, physician to King Charles I, who wrote a letter in answer to Thomas Merrick who was going to write a comprehensive British bird book (though he never actually went ahead with it), listing all of the birds in the Norfolk area. A renaissance man of many talents.

And outside of the natural world?

Joseph Haydn; whose good humour shines through his every composition.

CAROL AND TIM INSKIPP

Carol and Tim Inskipp have done much to raise the profile of Asia's birdlife and have pioneered many projects to highlight the importance of Nepal and its threatened birds in particular. They were both born in the 1940s.

INTERVIEWED BY KEITH BETTON

Where were you born and what was your first experience of birds?

Carol: I was born in Bishop Auckland in County Durham. I don't remember my first experience of birds, but my mother told me I was fascinated by birds even when a baby in the pram!

Tim: I was born in Hastings in Sussex. My first birding memory was being attacked by a Mute Swan in the local park, aged about five. This apparently did not put me off, and by the age of ten I was actively watching birds.

Who influenced your birdwatching in those days?

Carol: My early birdwatching was most influenced by lads from Hartlepool, Stockton and Middlesbrough who visited Hartlepool Headland and the Tees estuary almost every weekend. I was about sixteen and used to catch the train to Middlesbrough as early as I could on Saturday morning and then get a bus to Hartlepool, meeting up with them at the seawatching hide on the headland. I especially remember birdwatching with Tom Francis, and I learned a lot from him, and also from Edgar Gatenby, a very amiable local birder who was always about at Teesmouth.

Tim: My parents, both of whom had an interest in birds and used to take me out to local sites such as Pett Level and Rye. My dad took me to Dungeness on the back of his scooter in October 1955, and we stayed in the Bird Observatory, where Bert Axell was the warden. Seeing Great Grey Shrike and various birds in the hand that had been caught for ringing made a great impression on me and I was hooked for life.

How did your early birdwatching experiences develop – and did you look at other wildlife?

Carol: Initially my birdwatching was influenced by day trips to the northern dales in County Durham and North Yorkshire, also to the Northumberland coast. We had frequent family camping weekends to the dales and the Lake District. Every year we had a three-week camping holiday to different parts of the UK, often to the Scottish Highlands where I remember finding my first Osprey catching a fish on Loch Morlich in the Cairngorms. The Scottish holidays gave me the opportunity to find Crested Tit, Black Guillemot, Golden Eagle and Scottish Crossbill for myself, which I found very exciting at the time. Also especially memorable were two boating holidays on the Norfolk Broads, where I was over the moon to see a Marsh Harrier on Hickling Broad. My dear grandfather gave me his opera glasses (I can't imagine how I saw anything with those, but I was thrilled with them all the same) and my grandmother bought me my first bird book – *A Third Book of British Birds and their Nests*, one of the Ladybird series with a male Stonechat on the cover. However, my parents did not encourage my birdwatching and kept telling me I would grow out of it – but after I met Tim they gave up that idea.

Tim: Aged thirteen I did an exchange visit with a boy from Reading, John Hodgson, who was keen on plants (and who went on to become a professor at Sheffield) and we shared our enthusiasm and knowledge, with me learning a lot about plants and John about birds. This has remained a major interest for me till this day, including a long-term ongoing survey of the plants of Dungeness.

At an early age I also developed an interest in butterflies and other insects and I remember being thrilled with seeing a Large Tortoiseshell in the grounds of my primary school. I have since got into dragonflies, moths and almost anything else that pops up in front of me. When I started birding, binoculars were expensive and generally quite poor. My parents had a pair of 8×26 French binoculars, which I was able to share, but it was a few years before I could afford to buy my own.

What was your first telescope?

Carol: My first telescope was a Hertel and Reuss.

Tim: I saved up from doing a paper round to buy an ancient brass telescope, with four draw tubes and about four feet long when extended. No tripods then of course, so it meant finding a suitable post to rest it on and then trying to locate the bird in the very narrow field of view it sported. So small birds were not really possible and by default I concentrated on ducks and waders.

Did you join the Junior Bird Recorders Club or the RSPB?

Carol: I joined the JBRC when I was twelve or thirteen years old, and have been a member of the RSPB since the age of sixteen or seventeen.

Tim: I only joined the RSPB, but not until the late 1960s.

What about your time at school – did you go birding?

Carol: My time at school was uneventful and did nothing to encourage my passion for birds.

Tim: Birding whilst at Hastings Grammar School was constrained by an antiquated policy of having school on Saturday mornings, partly offset by having Wednesday afternoons off. In the late 1950s I regularly used to take a bus to Pett Level on Wednesday afternoon and use the sea-wall as a convenient scope perch, which got me familiar with all the common ducks and waders and the occasional scarcity such as Red-necked Grebe. I used to go through Alexandra Park between home and school, which in those days hosted regular Hawfinches and Lesser Spotted Woodpeckers. My parents took me on various trips: Cley in April 1956, Snowdonia and several trips to France stick in my memory.

In those days, who were your main birding companions?

Carol: One Easter I made my first visit to Spurn, where I stayed at the bird observatory for a week or so. Barry Spence was the warden. I remember meeting Barry Banson there and didn't meet him again until years later when he had retired and enjoyed sea-watching frequently at Dungeness. I remember he told me he couldn't set out for birding until he had had his morning coffee, which mystified me at the time, though I take the same view nowadays. There was snow that Easter, but we did have a fall of birds including my first Bluethroat. It was very cold and I felt sorry for some lads from Hartlepool (including Keith Redshaw) who were camping nearby so I lent them some observatory blankets and got into awful trouble about it, even though no-one was using them at the time. Although birding is still my main hobby nowadays, I have become very interested in other wildlife that flies, especially bats, butterflies, moths, dragonflies and bumblebees, and I record these as well, especially in Weardale in the north Pennines where I have a cottage.

Tim: I did not meet anyone else at school with an interest in birds so, apart from being taken out by my parents and the influence of my biology teacher, who was proficient in many aspects of natural history, I birded almost entirely on my own.

What about your early bird books?

Carol: My early bird books included the *Observer's Book of Birds* and the Collins *Field Guide to the Birds of Britain and Europe* by Peterson, Mountfort and Hollom.

Tim: I also started with the *Observer's Book of Birds* and the *A Bird Book for the Pocket* by Edmund Sandars (with awful plates), but my parents bought the Peterson, Mountfort and Hollom field guide soon after it first appeared and this made things so much easier for me in some ways (though in others, such as the large number of species to look through, it was a challenge to use).

Were you good at keeping notes?

Carol: Yes, I have always kept notes and still have all my notebooks – the first notes I made were in 1958, when I was ten years old, and described counting the numbers of Grey Partridge coveys that I could see on the weekly hour's bus ride to Sunderland to visit my grandparents.

Tim: I kept good notes in my early days but unfortunately have become lax about it in recent years.

To what extent were you a twitcher?

Carol: I became a twitcher after meeting Tim in 1972 and think I was until about fifteen years ago, although I still enjoy a twitch enormously when the opportunity arises.

Tim: I never had the resources or information to do any twitching whilst at school and university but in the late 1960s I met up with several enthusiastic birders, including Dave Holman, Paul Dukes and Pete Burness, and we travelled extensively around the UK. This continued in the 1970s when I met Carol, with regular visits to Scilly and weekend trips to many places. Subsequently we couldn't keep up the pace and twitches were reduced to birds we especially wanted to see and those within easy reach. The latest distant twitch we have made was for a Brünnich's Guillemot in Portland Harbour in 2013.

What have been your best birding finds in the UK?

Carol: My best find was a Black-winged Stilt at Beeston Rylands, Nottingham, in May 1973.

Tim: By chance my first Black-winged Stilt at Camber in September 1958 sticks in my memory as my first rare bird. Other pleasing finds were Little Bustard on St Agnes in October 1975, a Fea's or Zino's Petrel at Dungeness in October 1982 – the first for the UK, and an adult male Pallid Harrier seen twice in the summer in Cambridgeshire in 2011 – the first records for that county.

What about your career – how did it start and what was it like?

Carol: I was very lucky to go to university, first at Sheffield to study chemistry and then at Durham for an MSc in ecology. After that I worked as an analyst for the River Dove Water Board in Leicestershire for a year or so before joining a bird survey team on Foulness Island, Essex, studying potential impacts from bird strike on the Maplin Sands, a possible site for the third London airport. This was probably my most enjoyable job, going birding every day. Foulness Island was a fantastic site for Rough-legged Buzzards, Hen Harriers and Short-eared and Long-eared Owls. Sadly, our work was cut short as Maplin Sands was abandoned as an airport site for financial reasons. Then I had an unremarkable and quite unsuccessful four-and-a-half-year career as a chemistry and science teacher at comprehensive schools in Basildon, central London, Wanstead and Huntingdon. In between teaching jobs I joined Tim on two exciting trips, each for a few months, to India and Nepal. Apart from one year working for the RSPB in 1988, the rest of my working career has been as a freelance researcher and writer. For about twenty years I wrote environmental education materials.

Tim and I have co-authored bird books since 1985. After our first visit to Nepal together in 1977 we found that little was known about bird distribution in the country at that time. Tim had the idea of gathering as much information as we could and producing a bird distribution atlas for Nepal, and I was very happy to join him in this project – which culminated in our first book, *A Guide to the birds of Nepal*, in 1985. We are indebted to Mark Beaman, who encouraged us and convinced Christopher Helm that the book was worth publishing. We are also grateful to Christopher for believing in us and for encouraging us to write a second edition of the book in 1990. In 1988, Mark Cocker and I wrote a book on Brian Houghton Hodgson, a fascinating nineteenth-century naturalist who lived in Nepal for over twenty years. Other Nepal bird books I have written include *A Birdwatcher's Guide to Nepal*, published by Prion in 1988, and *Nepal's Forest Birds: their Status and Conservation*, published by the International Council for Bird Preservation (now BirdLife International) in 1989. Since 1998 I have co-authored a number of field guides with Richard Grimmett and Tim, including *Birds of the Indian Subcontinent* and various spin-off books including a *Birds of Nepal* field guide.

Since 2003 I have been researching and writing Nepal bird conservation books with Dr Hem Sagar Baral, a very knowledgeable Nepali birder I have known since 1986. We first co-authored *The State of Nepal's Birds 2004*, followed by an update in 2010; also the *Important Bird Areas of Nepal* in 2004, which I am currently updating. The largest research and writing project I have undertaken is the Nepal national Red Data Book, which is a detailed assessment of all the country's bird species, based on IUCN criteria. It was published as an online resource for free download in February 2016. The book is very much a team effort, and the contributors were mainly Nepalis; Hem and I led the team.

Tim: I qualified with a degree in botany from UCNW Bangor in 1966 but had no real ideas of a career then. I added an education qualification in 1968 but this only convinced me that teaching was not for me. I had a series of short-term jobs around this time, including at Colchester Natural History Museum, where I learned to skin birds, and an adventure centre in Wales, where I quickly became the caving expert because no-one else wanted to do it! Then I embarked on a mind-numbing year in a life insurance office in London, to save up enough money for a planned overland trip to India and Nepal.

After returning from India I had a year of more short-term jobs before taking on a two-and-a-half-year contract with the RSPB to study the importation of birds to the UK. This involved regular visits to Heathrow to see consignments of birds arriving, and many visits to zoos and bird dealers. It culminated in the production of a publication, *All Heaven in a Rage*, and was followed by another short-term contract leading to *Airborne Birds*. This experience led me to become the first employee of the newly fledged organisation TRAFFIC (Trade Records Analysis of Flora and Fauna in Commerce), which John A. Burton set up as an IUCN Specialist Group. This was based in Soho for several years before transferring to Cambridge to join other people working on IUCN Red Data Books. The organisation changed in structure and affinities several times, eventually becoming the UNEP World Conservation Monitoring Centre. I continued to work there on wildlife trade related matters, with the emphasis on CITES issues, until I retired in 2006. Since then I have had some consultancy work with WCMC, some proposed wind-farm surveys, and various botanical surveys, mostly unpaid.

When did you first go overseas birding?

Carol: My first overseas birding was with Tim when about ten of us went to the Camargue in a Bedford van in April 1973.

Tim: My first major trip was an overland journey to India and Nepal with Ron Brown, Peter Clement, Alan Greensmith, William Howard and Bob Jarman. We set out on 5 September 1970 in an ancient Bedford Dormobile that we bought for £60. It almost immediately packed up in Belgium, needing a new gearbox. Thankfully it kept going after that through Europe, Turkey, Iran, Afghanistan and Pakistan, arriving in India at the end of September. For the following five months we travelled in northern India, as far east as Assam, and in Nepal. In March 1971 we sold our vehicle in Kathmandu and went our separate ways – I headed down to Bombay. By this time the £200 I had started with was much diminished, so, to save money, I put my bag in the station left luggage place and joined many locals sleeping on the grass outside. I took a train up into the Western Ghats to see some hill birds but, unfortunately, strayed into the grounds of a hydroelectric facility and was promptly arrested. My captors spoke no English and held me for six hours waiting for an interpreter. Fortunately, I had no camera and I was able to persuade them by showing notes and sketches that I was an innocent birder. On leaving via the proper entrance there was a sign that I was liable to three years in prison! I booked a deck passage with the British India Steamship Company on a ship to Kuwait. This involved sleeping on the deck with 1,000+ Indians heading to the Middle East hoping for work. The voyage took two weeks, stopping to unload cargo at various places on the way and pick up more deck passengers. I was not well when we started, having picked up an irregular one-day fever (which persisted for about three years), bad diarrhoea and an extremely bad cold/flu. However, I survived and saw a few seabirds, including Jouanin's Petrels, Persian Shearwaters and Red-billed Tropicbirds. When we got to Kuwait I was not allowed to leave the boat because I did not have a visa for Kuwait (having been told by the British Embassy in Bombay that I didn't need one). They threatened to send me back to Bombay but eventually agreed to let me off if I took a taxi straight to the Iraq border. I had to change some money for this, virtually all of which was taken by the taxi driver, leaving me with £20 to get home! I managed to cross Iraq cheaply and crossed into southeast Turkey, where the local villagers all turned out to throw stones at me. I was arrested again and held for a few hours being shouted at in Turkish but eventually released. I crossed Turkey by bus and then, with a combination of hitching, buses and trains, I reached Ostend completely penniless. I had to beg round the ferry terminal to raise the crossing fee. So an eventful trip but very worthwhile with 600+ species of birds, many of which were new to me.

You are both strongly associated with Asian birding, but have you travelled widely elsewhere?

Carol: Other countries I have been birding are Australia, New Zealand, Costa Rica, Panama, Madagascar, Gambia, Zimbabwe, Zambia, Morocco, Thailand, India (ten times), Sri Lanka and many European countries, mainly with Tim.

Tim: My association with CITES and attendances at meetings of the Conference of the Parties and other meetings took me to a wide variety of countries: Costa Rica, India, Botswana, Argentina, Canada, Switzerland, Japan, USA, Zimbabwe, Kenya, Chile, Thailand, Netherlands, Qatar, Guatemala, Madagascar, New Zealand, Ghana and Vietnam. With Carol I have concentrated on visiting Nepal and the Indian subcontinent but have also been to Australia, Panama, Morocco, Zambia and many European countries.

When did you first visit Nepal? And why has it remained such a major focus in your life?

Carol: My first Nepal visit was in 1977, very soon after Tim and I married. He had regaled me with tales of that wonderful country ever since we met in the early 1970s. In 1977 Tim had an opportunity to study the wild bird trade in India for the organisation TRAFFIC, and I joined him for a four-and-a-half-month trip. We travelled to India on long-distance public buses, leaving in September – first taking a bus to Macedonia, then another to Istanbul where we enjoyed a few days watching migrating raptors. The next leg of our journey took us to Tehran – we set off with two drivers, but one was very sick and left us on the edge of Istanbul. In eastern Turkey the road was rough and unsurfaced and a stone smashed the front window of the bus. Our driver stopped the window from caving in by reaching out one hand to hold the window in place. He steered with one hand for a few hundred miles – amazing. He positioned his mirror so he could see me and kept trying to catch my eye. When we reached Tehran he tried to persuade me to return with him – he was a very handsome young Turk! In Tehran we camped for five days waiting for a bus to Herat in Afghanistan.

I found Afghanistan a fascinating country, especially as it was before the Russians invaded and the country was at peace. We spent about two weeks there, visiting Kabul, Kandahar and Jalalabad. We had a couple of great days birding in a desert en route between Kandahar and Kabul, where there was a large hotel built by the Russians. We were the only guests and it looked as if there had been no guests for a long time. Next morning, we were up early, birding in the nearby desert, but failed to take much notice that the hotel was adjacent to an army camp. That evening the army camp chief officer came to visit us and to our surprise spoke to us in perfect English, wanting to know what we were doing. He asked me the difference between a crow and a raven. After I answered he seemed satisfied that we were birders after all but warned us we had been very close to being arrested. From Jalalabad we took a bus to Peshawar and then a train to Delhi, and after that we visited different parts of northern and central India while Tim tried to find out what he could about the Indian wild bird trade.

After what was, for me, a rather harassing experience in India, we at last reached Nepal. The country far exceeded my expectations – the spectacular mountains, the amazing diversity of birdlife, beautiful forests and rivers, and the delightful humorous and kind nature of the Nepalis. I have visited the country twenty times, and if my finances allow I still aim to be visiting Nepal frequently in ten years' time. Although it has changed hugely since 1977, with far more roads, spreading urbanisation, increasing population, and much more traffic, I still love Nepal and am thrilled to return on every visit. Now I have many Nepali friends, and I enjoy visiting and staying with them as well. I have been very lucky to have trekked widely in Nepal, mainly with Tim. Our first trek was from Pokhara to Muktinath in Mustang along the Kali Gandaki valley in what is now the Annapurna Conservation Area – now a very popular trekking route. I have also trekked to the Annapurna Sanctuary, Langtang National Park in central Nepal; Sagarmatha (Everest) and Makalu Barun National Parks, Kanchenjunga Conservation Area and the Mai valley in the east, and Khaptad and Rara National Parks in the west. In 1982 Tim and I had a fantastic opportunity to make the first survey of the Bengal Florican, an endangered bustard, in Nepal's lowland grassland protected areas. My favourite birding places anywhere in the world are Nepal's Chitwan National Park and Koshi Tappu Wildlife Reserve, which I have visited many times – and Weardale in the north of England, close to where I grew up.

Tim: As I mentioned, my first visit to Nepal was in 1970–1971, when I was there for about three months altogether. Kathmandu was very quiet then with only three paved roads and only a few Ambassador taxis. We camped at Godavari for a week and were amazed by the sheer numbers and varieties of birds – one day there were an estimated 1000+ *Phylloscopus* warblers (Hume's and Lemon-rumped) on Phulchoki. I did a trek north to Helambu with Bob and Will but the extremely poor map we had led us on a non-existent trail on a high ridge. At dusk we had to drop down to camp by a lake far below, but the last ten metres was fairly sheer and, although we safely slid down, we could not return the same way! We spent three days looking for a way out and eventually managed to climb back up (otherwise we would still be there). We wanted to visit Pokhara but there was no drivable road from Kathmandu then. We had to return to India and find a way past broken bridges, and other obstacles, across the Gangetic plain to reach the border at Bhairawa. Our passports were stamped out of India but there was no border post on the Nepal side so we had to continue without. The Pokhara area was very attractive then with virtually no buildings by Phewa Tal and I had a successful time there, including finding White-tailed Eagle, Common Goldeneye and Reed Bunting – all new for the country. There were even fewer vehicles there and, whilst waiting for Bob and Will to return from a trek I acted as a taxi service, on one occasion carrying twenty-three people, including a pregnant woman who needed to get to hospital urgently. We went back to Kathmandu the same way and then went our separate ways, selling the now almost undrivable vehicle for £75.

When I returned with Carol in 1977 we decided it would be good to prepare an atlas of distribution for Nepal's birds, which would usefully complement the field guide: *Birds of Nepal* written by R.L. Fleming and his son, which was first published in 1976. With this in mind we returned for several more trips in the early 1980s, leading to our *Guide to the Birds of Nepal* in 1985. Not satisfied with that, we continued to visit and produced a second edition in 1991, followed by *Birds of Nepal*, a Helm field guide, in 2000. We both love the country and its birds so are still returning as often as we can. There are still plenty of areas we would like to visit and a few new birds to track down.

If I could introduce you to people you've never met – alive or dead – who would they be?

Carol: I would love to have met Brian Houghton Hodgson. It would be fascinating to meet Desirée Proud, who lived in Nepal in the late 1940s and 1950s and was a keen birdwatcher. She made several treks in what is now Langtang National Park and published several bird papers. Also Professor Jochen Martens, who did so much pioneering ornithological work in Nepal.

Tim: Allan Octavian Hume and William Edwin Brooks, who were ornithologists in India in the late 1800s.

Where is the best place you've ever been birdwatching?

Carol: It is hard to pick just one place – there are so many wonderful birding localities in Nepal.

Tim: Chitwan National Park, Nepal, where we once saw about 170 bird species in a day in 1982.

Where is the worst place you've ever been birdwatching?

Carol: Worst place I have been birding is Madeira – I didn't like the extensive tourist development on the island.

Tim: Parts of Myanmar where, because of extensive hunting, there are very few birds and, where there are some, they are excessively shy.

If you could go birding to one more place in your life that you've never been to, where would that be?

Carol: If I had a chance to go birding in one more place, I would choose the Himalayas of northeast India.

Tim: New Guinea.

What is your favourite bird group?

Carol: Owls.

Tim: Babblers.

What is your most wanted bird?

Carol: Cheer Pheasant.

Tim: Himalayan Quail.

Favourite piece of music?

Carol: I don't have one.

Tim: I don't really have a favourite either! I prefer to listen to something that's new to me.

Favourite film?

Carol: *Sense and Sensibility.*
Tim: Nothing comes to mind – sorry, I don't really have favourites.

Favourite TV show?

Carol: *Death in Paradise.*
Tim: Nothing for me.

Favourite non-bird book?

Carol: *The Illustrated History of the Countryside* by Oliver Rackham
Tim: *The Prophet* by Kahlil Gibran.

Favourite bird book?

Carol: The *Collins Bird Guide* by Killian Mullarney, Lars Svensson, Dan Zetterström and Peter J. Grant.
Tim: *The Birds of the Western Palearctic,* nine volumes by Stanley Cramp *et al.*

BARBARA YOUNG

*Baroness Young of Old Scone has worked in the health
service and the environment movement most of her life;
amongst other roles she has been chief executive of several
health authorities, the RSPB, the Environment Agency and
Diabetes UK. She sits in the House of Lords. She was born in
the 1940s.*

INTERVIEWED BY MARK AVERY

Where did you grow up?

I was born in a little village called Old Scone, in Perthshire, which
had forty-two inhabitants and was the feudal village of the Earl of
Mansfield. So everybody there worked for the Earl of Mansfield,
including my dad, who looked after his horses and did things like
driving the lorry, and cutting the cricket pitch and cutting the
grass at Scone Palace, and stuff like that. Dad was an engineer and
Mum was a mum. Both my parents were from the same village in
Perthshire, one village away from Old Scone.

I can see why you are Baroness Young of Old Scone.

Yes, but it was difficult getting the title. I had to get the permission
of my feudal overlord, the Earl of Mansfield, and he was thrilled to
bits actually – I thought he might be a bit sniffy about it. And also
we had to get the Scottish College of Heralds to agree as well as
the English College, and one day I was sitting in my office at RSPB
when I was told there was someone from Disney on the phone.

I said, 'What do you mean, someone from Disney?' and she said it was 'Lion King' – but it was Lyon King of Arms. He was a bit sniffy about it, but the Garter King of Arms, who was his opposite number in London, turned out to be an expert on South American birds so he was very helpful.

Long before you got your title, when you were a little girl, is that when you got into horses?

I rode before I walked – from eighteen months. Every waking hour, every waking hour! I used to leave at crack of dawn and head for the horses and come home at six for supper. Except of course when the Mansfield family came home, because they wanted to ride their horses – and I thought this was a bit much.

Was nature important to you at that time?

This was a very rural area and the Mansfield family had a large shooting estate up the glen too. And we used to decamp with them. So I did beating and bringing in deer on the ponies. I was born and brought up with gamekeepers and of course the Mansfield family owned the fishing on the Tay so we ate quite a lot of salmon. And pigeons – I was shocked when I left home to find out that you had to pay for pigeons – they were just vermin – we used to eat them and hang them up on the fence wires to scare away the others.

As well as eating birds – did you like them and watch them?

Birds were everywhere, and it never entered my head that you had to look at them. They were just always there. When you rode through grassland you'd be flushing waders all the time – Lapwings, Curlews and Snipe, all the time – and sometimes your horse would wobble at the knees and shy, and you'd fall off. But these days those are all improved grasslands and there's not a wader to be seen.

So you wouldn't have called yourself a birder then?

No, and I wouldn't have called myself interested in birds either. They were just part of life and not something to be especially interested in. Although when I won a school prize I asked for a book on nature and I wanted something very detailed that I could learn from – I read voraciously at that age, encyclopaedias, everything, and I just hoovered up facts. That's why I know quite a lot about horses – it's because I have fallen off lots of them and have read an awful lot.

What about school?

I walked to primary school in the next-door village and then I went to Perth Academy.

A girls' school?

No, it was co-ed and a grammar school where we were still streamed.

I bet you did well at school.

I did! I did really well. When we did our Highers I was fourth in the whole of Scotland.

And then on to university?

Yes, I did classics at Edinburgh.

Why classics?

I had this brilliant classics teacher called Bubbles who was gay in the days when you weren't allowed to talk about that. He was the choirmaster too, and I sang and he taught me to sing, and he was exotic and he lived with a set designer for Scottish Opera and he made us learn classics by reading all the dirty bits and I loved it. I still love classics. I'd love to go back and finish my PhD some time.

Did you do well at university too?

No, I didn't do as well, and I'll tell you why. I was having quite a good time but it's partly because my mother was convinced that Greek was a waste of space and so she wouldn't let me do it at school until pretty late on. She was a keen European and thought French was the answer, and so I did Greek Higher in about eight months and when I got to university there were guys there who'd started doing Greek at the age of about six – and so I was brilliant in the Latin and struggled like hell in the Greek. Although, to be honest, although I thought I was working quite hard, I have it on good authority from people who were around me that I wasn't working that hard.

Another thing was that I had a big falling out with my Greek professor, Beattie, over his interpretation of Aeschylus (he was the world expert!), and that didn't help. But he had a soft spot for me, and he was a keen birdwatcher, and we all got into trouble for using binoculars near the airbase at Corinth. It was the time of the Colonels. We only got out because the professor of archaeology at Athens noticed that we didn't seem to be around any more.

After university you got a job?

No I didn't. I was so busy trying not to fail my Greek exams that I didn't apply for anything, so I did a postgrad secretarial course. I was a low aspirer. I was engaged when I was at university. In fact, I was engaged twice when I was at university. This was long enough ago that I imagined that I would get married and have kids and not work. Well, none of that happened.

I took that secretarial course and I was bloody useless. And I applied for the graduate training scheme for the health service and they turned me down at the first interview, and I was, again, somewhat pissed off and thought, 'Nobody's turned me down for anything before and I'm not having this' – so I went into the health service as a secretary and after three months my boss said to me, 'You're absolutely hopeless at this, but we have a job as a communications/public affairs officer – would you like to do that?' So I did that for eighteen months and it was great fun. The day I started, my boss was on holiday and a lady on our patch had quins without fertility drugs and they all survived, the Bostock quins, and I handled all the media work for that. The world's press went wild and we negotiated with the local authority that they would give the Bostocks a bigger house.

After about eighteen months I decided I was fed up with talking about it and I wanted to do it, so I got a job as an assistant manager in a hospital in Glasgow. I was living with someone at the time and we were breaking up in a rather messy fashion after five years so I moved to Glasgow. The two blokes I worked for were Bert and Dennis, and they were brilliant. They were Jack-the-Lads in lights, but they were really good to me and I had to stop them doing all sorts of daft things. We appointed a secretary once to work for us. These two had seen her walking across the car park and their eyes were popping out, so I knew that we were going to appoint her whatever happened. And she was absolutely brilliant and gave them short shrift, but I've lost touch with her over the years, which is a shame because I have this annual get-together with all the PAs who have worked for me and she's the only one I can never track down – she's probably back in France.

Then I was promoted and promoted and I was running about seven hospitals in Glasgow. And I really loved that because doctors are really great. I was about thirty-one.

What did you do in the health service then?

Well, you shape the strategy, manage the staff, plan what they are going to do, you manage the budgets, you open things and you close things. I've built several multi-million-pound buildings in my time. We once bought the hotel next door because we needed more space and we put all the old dears who didn't need heavy-duty nursing in there – and they had a really nice time and that was much, much cheaper than paying for them to be bed-blocking. You had to make sure that we treated patients to a high quality. You had to make sure that they didn't kill anybody.

I moved to London after a while because there was no way that I could get promoted any further in Scotland. So I decided I'd go to London. I applied for lots of jobs and kept coming second for them. So then a friend finagled me a job doing strategy and planning for St Thomas' Hospital, and it was quite different from Scotland. I was working with the London School of Economics and KPMG, and we had our three-day strategy session in Leeds Castle. The trustees threw money at everything. I took ages to settle but I finally did.

That role came to an end when we completed the strategy, and

I started applying for deputy jobs in health districts. I was offered two at once, Westminster and St Mary's. I chose St Mary's, and, to be honest, it was the best thing I ever did. We did lots of terrific stuff. And I was active in the Institute of Health Service Management, which I eventually chaired and then became president of, so I was campaigning actively for the NHS. I was doing media work, and lectures all round the world, and meeting lots of politicians and ministers. So I was becoming a bit of a rent-a-mouth for the NHS – you know, you know me!

And then the health service was restructured and they decided I ought to run a health district of my own – and it was Haringey, and it was the punishment cell. The medical leadership was non-existent. But eventually I found one medic who seemed to have a bit more about him, and I kind of encouraged him to lead the docs until about nine months later the regional press officer phoned me up and told me to buy a copy of the *News of the World*. I asked why, but he said just go and buy one, you'll see. And this medic I'd chosen wasn't just on the front page of the *News of the World* but on four pages inside having had wild sex parties, the bastard! This was at the time of the riots in Broadwater Farm Estate where the guy got decapitated. And Bernie Grant was on my authority – it was steamy. We were picketed and I got my nose broken by pickets.

Then Ken Clarke reorganised the NHS and split the providers away from the purchasers and I looked at these jobs and thought I don't like the look of these because I like doing both. So, forget it! The nice thing had been that you could look at a population, and with it decide what it needed, and then deliver it and get the right sort of outcome.

I knew I was going to go. I'd done twenty years in the NHS, but I didn't do anything about it for a couple of years. I was on holiday with friends, and I saw the RSPB job advert and I told my friends that was what I was going to do. I didn't know much about the RSPB other than that it was a trusted brand.

But at that time, in our office in Paddington, my PA, Debra, who came to the RSPB with me, and I had started watching the birds on the towpath of the Paddington Basin. Debra had a bird book and a pair of binoculars and we would watch the Canada Geese and the ducks, I guess Mallards, going up and down, and then the RSPB advert appeared in the papers. It shows our ignorance that we were getting really excited about Canada Geese and Mallards.

About three weeks after I got back I had a drink with my friend and he asked how the RSPB job was coming along and I said, 'Fine, fine' – so I rushed back and found that the closing date had passed three days ago. I phoned up the head-hunters and told them, 'You can't afford me, I know absolutely bugger-all about birds and conservation, but these are the things that I can do.' I dropped my salary by £40,000 to go to the RSPB but, what the hell, I wanted the job.

What did you make of the RSPB when you arrived?

I kind of worried that the charity sector might be all threadbare carpets and ancient desks, but of course the RSPB wasn't like that at all. The thing that was different – well, lots of things were different, of course – but the thing that caught me was that the NHS was always under pressure and the limits of tolerance were very narrow, which meant that if anything went wrong at all then there was a total shambles. Whereas the RSPB believed in doing things well and resourcing them properly and so most of the time things went pretty well. And it was just like chalk and cheese, and it made me realise why the NHS was stumbling along making a pig's ear out of lots of things.

What about the people?

The thing that struck me about the RSPB, and I think about environmentalists generally, is that they are different from the charity sector. The charity sector can be woolly and fluffy and 'I'm here for the beneficiaries' – and that hides a multitude of sins and that drives me bonkers, whereas if you look at RSPB, Birdlife or the BTO, most of the folk who work there would do that work even if the organisation didn't exist. They are really committed to doing those things. I always used to say in the NHS that 'Management is not there to get in the way, it's there to help you do what you want to do. And surgeons want to cut people, so I'm there to help them cut people.' And that's what it was like at the RSPB. The job was to make sure there wasn't anything structurally or in terms of resource getting in the way of people who were really knowledge-able and really wanted to do the things they wanted to do.

I was very junior when you arrived at RSPB but obviously we were very interested to see what this woman, whom none of us had heard of, was going to do. I might have remembered this wrong, but Ian Prestt was the Director General, which was like the Chief Exec, and you came in as a sort of Managing Director?

Well, no, I have a terrible confession to make about that. I was a bit worried because the plan was that Ian would stay as Director General and I would come in as Chief Executive – now, what the hell is that all about? And I said to the head-hunters, 'I'm out of here,' and they said, 'Hang on a minute. Ian is ready to take a step back and you'd be the boss.' So I came and saw Ian and said, 'Look Ian, if I am going to do this job then I'm going to have to be able to do it properly and I'll have to be the boss. How do you feel about that?' And he said, 'That's OK.' I hate to say this, but I got appointed in the October, and I was going to arrive in January, and Ian had his stroke between Christmas and the New Year. To this day, you know, it was horrible. And it was horrible for Ian. But it was the saving of me because I think it would have been really difficult. And so I was always Chief Executive right from the start.

And Ian became President. He was kind of kicked upstairs.

And he was sweet. He was a real sweetheart. He was a lovely man.

Surely this was a dead easy job compared with the huge jobs you had been doing in the NHS?

It was quite small. But it was quite different. For example, I hadn't had to bring in the money before.

In the health service there was always this issue about controlling the spend because there were all these demands and patients piling up against the doors and I remember crawling over the figures at the RSPB and the Finance Director, David Gordon said to me that I didn't have to worry about that – the difficulty would be spending the money. That was different.

I did know how to run an organisation, and that was regarded as black magic for the first year or so. And people went, 'Ooh! She does all these things we've never seen before,' and all I was doing was bog-standard management things. That bought me time until I swotted up.

You sent us all off, the senior and middle managers, on training courses that became known as 'sheep dip' because they got rid of all our parasites and we all came out smelling the same.

Yes, that was what it was supposed to do – not so much get rid of parasites as make sure people had some sort of common language and broad direction.

Well, they were very good. Mainly because we met lots of our colleagues that we didn't normally meet, and got drunk with them, and realised they were actually quite nice, and then it was more difficult to 'hate' marketing folk because one actually knew who they were.

I've got to ask you whether this story was true, and we all believed it was true, and wanted it to be true, and so spread it around for all we were worth, and that is that in one of your meetings with your directors you picked up your papers, threw them up in the air, and said 'You're all effing idiots,' and walked out.

Well, it was worse than that if anything. We were having this meeting, and a more junior member of staff was making a presentation which was quite sensible but it wasn't brilliantly done and the paper wasn't great, and some of them, the difficult boys in the room, gave the presenter a hard time. And they weren't doing it because the proposition was bad but just to be difficult. And I was beginning to see red on this. So what I said was, 'I don't want to work here with a bunch of effing amateurs' – and I picked up all the papers, theirs as well as mine, and threw them out of the window towards the pond. And then I said, 'This meeting is finished. We will make decisions about the remainder of the items on this agenda and I will tell you what they are.' Didn't have to do that again.

At the RSPB you became a Labour peer, and that caused a bit of a stink. Had you been a Labour Party supporter for ever? Surely the Earl of Mansfield would have been shocked?

I kind of got politicised early on and I always knew I was Labour. My parents were Conservative. In fact, Mum was a floating voter – she voted SNP sometimes. I helped campaign for the abortion law reform in 1968 and then I didn't do politics – well, that isn't true, because we bunged out Malcolm Muggeridge as Rector of Edinburgh University because he said that the student health centre shouldn't hand out the Pill, and I wasn't having that, so we gave him the boot. That was quite fun.

In Glasgow you couldn't really be anything other than Labour, despite the fact that I lived in Hillhead, which was Roy Jenkins's seat as a Social Democrat. In the only election I voted in in Hillhead I couldn't think what to do so I drove my car into the back of his, quite deliberately. I never said it was deliberate, it was a very steep hill! But I thought if I could keep him off the campaign trail, even for a while, we might get a Labour MP.

At that point I got in tow with Donald Dewar, who was the MP for part of my patch, and I was very fond of Donald. And on a Saturday I used to sit in my mate's parents' kitchen – his dad was General Secretary of the Scottish Trades Union Congress – drinking cups of tea, with the political elite of Scotland passing through and generally chewing the fat. And it was great, I loved it. So you just slide into it.

I've never been a great constituency Labour Party person. I've tended to go straight to the top.

What a surprise!

And then in London, Robin Cook, Gordon Brown and Donald Dewar were all active. Tony Blair was deeply envious of the RSPB membership of a million. Margaret Jay was on my health authority, and because I was active in the NHS I got to know an awful lot of politicians. David Willetts was on my health authority for four years.

But when you became a Labour peer at the RSPB it wasn't very popular with much of the establishment.

No, half of the vice-presidents had a hissy fit. Wendy Nicol was brilliant on that and John Moran was…

…not thrilled?

…not thrilled. But Wendy had a meeting of them all and coaxed them into line.

John Lawton was Chair of RSPB Council, wasn't he?

Yes, and John was extremely helpful and I made a deal that I would go in a year or so, when I would have done eight years as Chief Executive. Gathorne Cranbrook was the Chair of English Nature and wanted to stand down so I went to see him and, I don't know how I had the brass neck, but I said that I couldn't afford to because that wouldn't pay very much. He was on the Board of Anglian Water and he was very sweet, and said he'd stand down from that and put a word in for me there too. He was so keen to get me to EN!

So, I had the House of Lords, Anglian Water and English Nature – and then out of the blue I was asked to be Vice-Chair of the BBC. And I thought that I'd love to but I just didn't have the time and I'd committed myself to so many other things, so I said no. But I told the Chief Whip in the House of Lords that I had given it up so I could be in the House and vote and he told me I was mad – and I really wanted the BBC role – so I got back to them and said, 'You know I said "no". Can we make that a "yes", please?'

And I loved the BBC job. I loved all of them actually, but I didn't quite love being portfolio. I was still in my forties and I kind of felt I didn't belong anywhere, even though each one of these jobs was really interesting. And I was horrendously busy.

But I still liked running big organisations, so eventually I moved on to the Environment Agency and now I have just left Diabetes UK to take up the role of Chair of the Woodland Trust.

Do you think that nature conservation is a political issue?

Oh yes! If you think about it, nature conservation is about a whole load of issues which are political, like the planning system, agriculture, infrastructure, transport, water, climate change – like Badgers! – they are all incredibly political.

Nature conservation needs enough people in the right places with political judgement and contacts to make a difference. I can give you an example from Diabetes UK: we had stopped being a campaigning organisation, God knows why! For me, campaigning is applying everything to the subject. We spend about £40 million a year on helping people with diabetes and the NHS spends £10 billion. So if we get leverage on that £10 billion we can make a difference. And it's similar with Common Agricultural Policy budgets – if you can change how CAP money is spent, even just a little bit, it makes a huge difference right across Europe.

As a fellow Labour Party member, I sometimes despair a bit about…

Well, I despair about you if you are going to vote for Jeremy Corbyn to be leader…

Well, I have, and by the time this book comes out we'll all be able to see whether that was a complete disaster or not!

…but the Labour Party always seems to be full of urban intellectuals who don't really seem to get nature and wildlife.

Well, I despair about it as well. The only thing is that you can make progress round the side because Labour is committed on climate change (and Ed Miliband was a real star on that) and the sense of place. They are committed to the idea that a planning system is an important democratic issue. They are pretty committed to natural resource management: they say the right things about water provided you keep the cost of bills down for everyone. And they quite like Badgers, so that's a good thing, and they don't like landowners, so that's a good thing. And they're not like some of the Tories who think that anything with a beak or claws is the Antichrist, so they have quite a lot going for them even though they aren't naturally 'Wow! Birds! Wow! Conservation! Wow! Countryside!' – but you can get them into the right place. I don't mind if people do the right things for the wrong reasons, provided they do them.

Tell me about the House of Lords. Do you do any good there?

Yes, we do a lot of good. There is no doubt about it, we do improve legislation because there are a lot of people there who know what they are talking about.

There are lots of things wrong with it too and I would support reform. We should have a retirement age, and there are far too many of us. Cameron has made it worse by putting people in who have no experience of parliament and immediately become ministers – they don't have a clue and then they often get bored.

Sometimes we put our foot down. I've got lots of good things through the House of Lords: the climate adaption legislation; schools having a statutory requirement to look after children with long-term conditions; a lot of changes to the Countryside and Rights of Way Act; the government to change what it was going to do on fracking (even though they then changed it back again). If you put your mind to it then you can actually make things happen.

And in the Lords we work together across party lines a lot. That's fun too.

You're the president or vice-president of just about every wildlife organisation out there. Do you think the NGOs are doing a good job as a gang? It certainly doesn't look as though we are winning.

I must admit I am a bit out of touch and I'm planning to work myself back in to these things. But I kind of get the feeling that we are less hungry then we were. We've become a bit domesticated. I think the NGOs need to do more to nail the government as well as work with them.

And that's my aim over the next five years!

There are many remarkable things about you, Barbara, but a fairly remarkable thing is that you came into the RSPB not knowing much about birds or nature conservation, but ever since you have stuck with it. And you have lots of chums, like me, who are birders, and you are a member of the Rare Bird Club, and you go off on birding trips…

…I even do a bit of Bedfordshire birding.

…so you seem to have the bug. Is it the people or the birds? Is it because we are all so lovely?

No – it's the birds. The people are lovely as well, but it's the birds. And I got the conservation bug very quickly – how could one not? But the birds and the wildlife and habitats are fascinating and they're beautiful. And at the RSPB and at BirdLife, every time you went out, you went out with the world expert – so I learned so much. People were so generous with their knowledge and their time. I remember being at Leighton Moss with John Wilson and he would walk into the hides and act as though he had never seen these things before, and yet he had done every day for about thirty years.

Colin Bibby once took me into a reed bed and we sat there all day and lived on reed aphids so that we could empathise with Reed Warblers. And in the RSPB we had 1,000 people like that and this other great family of people across the world in BirdLife. The people were great, and are great, but the things they showed you were fascinating.

Where have you been recently?

I went to South Georgia for five weeks

Not many people have been there.

And I've been to parts of South Georgia that even the people who have been to South Georgia haven't been to. I'm President of the South Georgia Heritage Trust and we've been eradicating rats on South Georgia for eight years now. We think South Georgia is now probably rat-free. And the team were just stunningly good, stunningly good.

People like Tony Martin?

Tony was a perfect project leader because he had so many experts, many of whom had led projects of their own, and you kind of thought there might be a clash of egos but there wasn't an ego in sight. It was hazardous, it was important, it was time-constrained, and they came across all sorts of blockages and disasters (two of the three helicopters were lost on day 2) – and they were just so good. Tony was just brilliant.

Do you have a favourite bird?

I used to say the Dotterel – it has a great lifestyle with the females in charge and they lay a clutch of eggs and then have a go with someone else or have a break on the continent or whatever. And I like the Lappet-faced Vulture because it is so plug-ugly. But actually the bird that I've been most knocked out by, I think, is the Giant Ibis – because it's like a leg of lamb on wings. There's no way that it should be able to fly if you're that shape but it does! And we yomped a very long way to get it so we had to work for it.

I like penguins too. If you sit in a penguin colony they come up and look at you and chat away to you. And they stare at you and they do a little jiggle, and if you do a little jiggle back they have another little jiggle back at you and they are always really interested in what you are doing.

That's how you think life ought to be, isn't it? People come up to you and do a little jiggle for you?

Yes! And mostly they do.

Favourite TV?

I hardly ever watch TV, and that was always a laugh when I was at the BBC. They used to send me a furniture van every Saturday morning at 8 am full of tapes and CDs and videos of things that I had to watch and know about, so watching TV in the daytime became OK for me – it was no longer sin, it was homework.

I like rather abstruse things on BBC4, like the *History of the Tractor* or *Lead Poisoning in the Middle Ages*.

Favourite film?

Field of Dreams: build it and they will come – that's my motto. And apart from anything else Kevin Costner is delicious to look at.

Favourite music?

Practically everything. I love jazz. I love Wagner. I'm very fond of Bach but I like Clapton and I always had a dead pash for Mark Knopfler [Dire Straits]. He was just drop-dead gorgeous. He used to live round the corner from one of my hospitals and I had this standing instruction that if ever he were to come in with one of his kids I was to be phoned – he never did, though.

Favourite book?

I used to like *The Leopard* by Giuseppe di Lampedusa, which is the story of the last days of the nobility in Italy when Garibaldi was unifying Italy. It's a really good story and there is a quite reasonable film of it with Gregory Peck.

BILL ODDIE

Bill Oddie is best known to many as a musician and comedian on television, but to many people he is Britain's best-known birdwatcher. He was born in the 1940s.

INTERVIEWED BY KEITH BETTON

At what point do you remember becoming interested in birds?
It was an offshoot to juvenile delinquency because every schoolboy of my generation collected birds' eggs, and by the time I was five or six, so did I. That doesn't mean that just because you had a collection you knew anything about it, and there was a very clear demarcation between the people who just ragged a nest – effectively destroying it – and those who were more interested in the eggs. So I definitely got into birds via egg collecting – as was often the case in those days. The trouble was that there was nothing much else to do, there was no TV and we were out playing in our spare time. So it was almost impossible not to know about birds at least a little bit. One of my seminal moments was finding a Pheasant's nest which had been deserted and I decided to blow the eggs. They were already rotten and by accident I sucked instead of blowing and I threw up all over them! In a way I was quite happy that the experiment had gone wrong, and I never collected eggs ever again. So by the age of seven or eight I had equipped myself with a good knowledge of where birds nested and the calls and songs they made.

Had you moved to the Midlands by this time?

We moved to Birmingham when I was about seven, and by the time I'd got there I would have definitely called myself a birdwatcher. I knew enough about it to say to my dad that I needed a pair of binoculars, and I took no chances – I went to an optical shop in town and asked for advice on what brand to buy. I was advised to buy Barr and Stroud 8×30 binoculars, and I had them for years.

Where was your main birdwatching haunt at that time?

It was Bartley Reservoir, and I spent much of my time avoiding being caught by the water bailiff. The reservoir is still in use and appears to be rather better than it was all those years ago. I spent hours and hours there during the school holidays, and if you look at my old notebooks they show that I was there from 6 am to 7 pm some days. Amazingly I hardly ever saw anything there despite walking around it endlessly!

Who influenced your birdwatching in those days?

I often walked around the reservoir in the company of George Evans, who went on to be the warden at Bardsey Island Bird Observatory in the late 1960s. He was ex-RAF and was an excellent tutor. We became really good friends. I went back there about ten years ago to make a short film for the BBC on my early birding days. It was still as bleak as ever, but having said that, I do remember getting hooked on visible migration when I was a kid. George and I would just sit on the reservoir bank for hours watching things fly over. I can remember quite clearly large numbers of Skylarks passing over in October – something you just don't see these days. George was convinced that there must be some passage of wading birds each year, and he was right. We put in a lot of effort and saw most of the common waders flying over. Migration watching is still the thing that excites me most, and I love to be in places where birds pass through. On a good day there is simply nothing like it.

What about holidays?

I never went on any holidays with my parents. My mother wasn't around and my dad never went away and so I was the one to go away. I was at King Edward's School in Birmingham. There were a few other birdwatchers there and that was why I first went to Dungeness, because one of the boys had been there already. For many years there would be a trip to somewhere – including Fair Isle several times and Out Skerries, and also Lundy.

When you went to Cambridge University, were you involved in local birding there?

I wasn't involved in the local birding scene there at all, and to this day I don't really know why not. The only place I knew nearby was Cambridge Sewage Farm and I did go there, and saw a few birds. On another occasion I got on my bike and cycled to the Ouse Washes. I was aware of the Cambridge Bird Club but I guess I didn't really want to go with a group. I have actually met quite a few of them since – such as Clive Minton, who was a frantically mad ringer even then. I never knew him at Cambridge but many years later met him in Australia when he was trying to ring a flock of waders. We went out in some kind of punt towards a distant island and the engine broke down on the way.

So I've never really been a club person, and I did not go to the West Midland Bird Club either. However, I remember contacting the club recorder in the early 1960s to report a Ferruginous Duck at Gailey Reservoir, so he went to check it out, and to his surprise it really was one!

I guess after university your life became incredibly busy. What happened to your birding?

One of the best aspects of having had birding in my life is that I continued to do that and fit it in to whatever schedule I had. So at the beginning of the year I would pencil in a week off in May and October but wherever I was I would be nipping out to go birding. And I still do that to a certain extent. In fact, some of my best days have been like that. I once had a day in spring in New York in Central Park after we had just finished a tour in America doing theatre shows. In every place we had been I'd managed to do a bit of birding for an hour or so – and I was never late for a show. When we started filming *The Goodies* we would always go away for two weeks to film outside sequences. I managed to influence the producer about where we would go. The most productive of all of those shoots was at Portland Bill in late September 1979. That included the famous long-staying Yellow-billed Cuckoo. I also got to see Red-throated Pipit and Common Rosefinch. A few years earlier, in 1973, we were filming at Portland and as we were filming an Alpine Swift actually flew over. I remember shouting 'Bloody hell, there's an Alpine Swift!' Thankfully it was silent filming. I told the unit manager that he might get a message from the bird observatory if something rare was found. One day he came and said that the observatory staff had been in touch to say there was a Cheese Sandwich Warbler showing well. I don't know how he came up with that, because it was a Pallas's Warbler.

Did you have time to be a twitcher?

Not really. Almost every time I give an interview they describe me as a twitcher and I have to correct them. My usual explanation is that all sprinters are athletes, but not all athletes are sprinters.

Back in 1980 I remember reading in *Bill Oddie's Little Black Bird Book* all the phrases used by twitchers, including the phrase 'cosmic mind f!!!er' to describe a mega rarity. Did you make that term up?

I had genuinely heard somebody say it in a hide somewhere. I don't think it was in common parlance at the time, but publishing it gave it a new status – or at least I like to think so!

I went through *British Birds* rarity reports to see how many good birds you found back then. In 1974 you found a total of eight rarities on the Scillies between 6 and 16 September!

It's all about timing and where you go. You read that a person found a rarity, but it could have been anybody – they just happened to pick the right place at the right time.

You did a lot of birding with Andy Lowe.

Yes, I did quite a lot with Andrew. We went to the Out Skerries, off the east coast of Shetland, when nobody else was going, so if there was anything good there we found it. Andrew and I had gone up to Shetland, having been to Fair Isle a couple of times, and thought nobody watched Shetland very much. More fool us as it turned out – not knowing where to start, we drove round and round for three days in pouring rain and saw absolutely nothing. Andrew said he knew Brian Marshall who was a doctor up there so we decided to give him a ring to see if he could get us out to the Skerries. We went across in fear of our lives in a rickety old fishing boat. We basically had the place to ourselves and subsequently went there several times. In 1983 I even took my wife and daughter when she was about two years old and we caught a Pallas's Grasshopper Warbler. We usually went in May – and other good birds we found in the late 1970s included Collared Flycatcher, Short-toed Lark, Thrush Nightingale, Great Reed Warbler, Snowy Owl, Lesser Grey Shrike, Red-footed Falcon and Rustic Bunting. I particularly remember the Collared Flycatcher, as a local resident had told us about a 'wee black and white bird' in her garden. We had suspected it was going to be a Pied Wagtail – but it proves that you have to check.

In the Scillies you seem to prefer Bryher above all the other islands. Why is that?

I like the smallness of Bryher. When the boat goes off at 4 pm you have the place to yourself. I have been reasonably lucky there. I seem to have a golden touch with Solitary Sandpipers, finding one there and two others on Tresco with Robin Hemming and Andy Lowe. Other lucky finds on Bryher were Citrine Wagtail, Great Spotted Cuckoo, Red-throated, Richard's and Tawny Pipits, Siberian Stonechat and Black-eared Wheatear.

What about overseas birding? Are there any trips that are particularly memorable?

In 1979 Richard Porter was talking about going to India and I asked if I could go with them as I'd never really been abroad, apart from having worked in America. It was a complete revelation. I absolutely loved it and wondered why I hadn't done it years ago. So we went to India, Nepal and Thailand in consecutive years. In 1983 the BBC made a TV programme called *Oddie in Paradise* in Papua New Guinea. I went with Laura – it was technically our honeymoon, paid for by the BBC! We also went to the Seychelles several times. For some strange reason it became a bit of a regular thing. The Seychelles Tourist Board phoned my agent and asked if I'd go and write an article. Then for the second trip a friend of ours was doing a documentary. The third time I nearly drowned. I hate water, but I was pottering around at the edge of a reef on Bird Island when the tide started to come in and I tried to get in but there were some currents against me. I was waving and shouting, and everyone on the beach thought I was having a lovely time and waved and shouted back.

So where were you birding when you got a bit busier?

I'd lived in more or less the same place for a very long time. Because I was successful in the 1970s I was able to buy a very nice house, and my back window overlooked the first pond on Hampstead Heath. I actually saw birds in the garden, but in a strange way it never really struck me that the Heath itself would be worth looking at. In the garden I saw Pied Flycatcher, Wood Warbler and Redstart, but I never really made the connection between them and the Heath. Then I started going to the Brent Reservoir quite a lot, and if you'd asked me then, I'd probably have said that was my local patch. I met Mark Hardwick at Brent Reservoir, and he watched the Heath avidly with Peter Dickinson, and he asked why I didn't, as they had no-one else to watch it and he could do with some help. So I switched from Brent Reservoir to Hampstead Heath and used to go there two or three times a week and still do. I have to confess a bit of possessiveness, as there's now potentially three or four other birders there. But I've now reached the age when I need a bit of help with top-range calls – for example, I have real trouble hearing Meadow Pipit.

You spent several years on the RSPB Council. What did you think about that?

It was when the RSPB was growing very fast, during the time when Barbara Young was Chief Executive and Julian Pettifer was President. I enjoyed being on Council but didn't enjoy meetings very much. And nothing's changed for me in that, not just regarding the RSPB – all NGOs in conservation speak in jargon and you can't understand the majority of it. At one of the first Council meetings I went to I stopped the meeting and told them I hadn't understood a word anyone had said so far. And it turned out that everyone else felt the same!

What do you think you did for the RSPB?

In a way, the best thing was when I left the Council. I talked to one of their press officers and he suggested that in some ways it would be better if I wasn't inside the RSPB because mentioning anything vaguely political or contentious was impossible. But if I wasn't on the RSPB Council I could say what I liked, so we had a tacit arrangement that if there was somebody or something that they would love to have had a go at, they would ring me and I could be controversial and name names.

Not many people in the world of conservation get recognised, but you got the OBE for your services to wildlife conservation in 2003.

I don't think it's any secret that I'm not a royalist, so being recognised by apparently the same system that gives OBEs to people who polish the corgis and that sort of thing really didn't mean a thing. I've worn it once, at a fancy dress party. And now, thirteen years on, if by any chance the next stage honour came along I think I'd probably turn it down. I did enjoy the experience of going to Buckingham Palace on the day, and the people who got honours for 'normal' things were great – and I was most gratified by the fact that nobody was taking it seriously and we were all laughing and giggling. There was very little reverence. And there are actually no dress rules – it doesn't even say 'no jeans' – so I wore a very colourful shirt. The Queen didn't say very much to me – something like 'It must be very nice making a job out of your interest' – and I replied that she'd made a film about the corgis which must have been fun too.

In 1980 *Bill Oddie's Little Black Bird Book* came out – which for me, and most of my birding mates, is still the one that made the mark. Do you feel that?

I do. The publisher wanted me to write a bird book, and in those days there wasn't much variety. I was at Portland one day, there was a bit of a twitch going on for a Red-throated Pipit – I was on top of the hill looking down on these little 'clans' of birdwatchers and this twitch was the classic example of the frantic activity going on, so that was the moment I realised that the book wasn't going to be about birds but about birdwatchers. And I simply wrote down what I'd seen and observed. I kind of lumbered myself a little bit because people assumed that was what I did, but I didn't really, it was just observation of others.

Do you find writing easy?

Once I get going. I do enjoy it, and once you've written a bird book with an angle you can't do the same thing again, so virtually everything after that becomes a pseudo-autobiography. I hope I'm coming up with new stories but I think I'm also getting a little bit more serious and contentious in my old age.

You've changed to be a little more campaigning in style – what caused that?

That also goes with not just being a birder. Without doubt one of the biggest pluses of actually making the TV series over the years was spreading from just birds into all wildlife and doing something new to yourself so that you're genuinely enthusiastic about it. There's a limit to how enthusiastic you can get about a bunch of gannets diving when you've seen it so many times. I genuinely love the items about wood ants or bush crickets.

Let's talk about birds on TV. Back in 1994 you presented a programme called *Bird in the Nest*. It was quite revolutionary, wasn't it?

Yes – I did that with Peter Holden of the RSPB. We did it for two years and it was exactly the same as *Springwatch* in many ways but purely about birds and was as live as they come. Peter and I had to ad-lib an awful lot, for instance if the news or the weather forecast was under-running they'd come over to us for an extra couple of minutes. It was quite pioneering and I'm amazed that it took them ten years to do it again as *Springwatch*. Peter is one of the worst gigglers – there was one instance about Robins nesting in an outside toilet and we discussed what would happen if for its first flight one of the chicks went straight down the pan. We went into the realms of fantasy about covering it with cellophane and the young bouncing off it – I did a series of drawings of Robin babies bouncing off a trampoline. All stupid stuff but great fun!

Springwatch and Autumnwatch were born from that then?

Somebody did suggest that we could do something like *Bird in the Nest* every year, but that was rejected, then *Bugged* came along, then *Wild in Your Garden*, and then *Springwatch* in 2005 which was a very good marketing title. The cooperation, planning, efficiency, technicality and all the people involved are undoubtedly things I'll never forget. It was a wonderful example of the sort of programme that only the BBC would ever make.

Was there competition between you and Kate Humble?

I got on her nerves terribly and then we figured out that it was because she had an earpiece listening to what was being said in the control room and (by choice) I didn't. So for instance she was being told to wrap up a sequence and cutting right across me. We were very fond of each other and got on extremely well, and it was just public conjecture that we hated each other. Kate and I used to be outrageous in rehearsal – we were always swearing and being purposely profane. We got our comeuppance once when we were told that there had been some Brownies in the control room who'd heard it all. We were very responsible when it came to doing the actual show. There were occasions during the last series when I was getting quite hyper due to the bipolar situation – it probably wasn't obvious to most people but I could be quite abrasive. During the infamous interview with Bill Turnbull I couldn't hear a thing and I was ad-libbing. My comment that he'd been reading the *Daily Mail* probably lost me a few points with the BBC.

I was disappointed when my involvement in *Springwatch* came to an end. *Britain Goes Wild* to me was the actual template, it was really great, there was a lot of freedom and taking of risks.

Those born in the 1960s remember you for *The Goodies*, and those born in the 1980s or 1990s remember *Springwatch*. When you meet people, does it annoy you that they think back to those but do not think about what else you've done?

Actually it doesn't bother me at all – people remember what they remember, it depends on their age.

Looking back on your TV and radio career, which makes you proudest?

One of my criteria for job satisfaction is to do shows which all the team – including technical, cameras etc. – enjoy doing and look upon as a challenge. That applied both to *The Goodies* and to the wildlife programmes, because they weren't easy to do and the special effects were complicated. To have people say 'I loved working on your show, I hope I can do the next series' is big job satisfaction. I also like it when people say it was part of their childhood or they loved the programme.

Do you enjoy watching episodes of *The Goodies*?

Some – I don't make a habit of it but quite often I have to, for example if I'm putting together a few clips for a talk.

Did you ever work with the Monty Python team?

Yes, and no. I'd been at Cambridge with several of them. John Cleese, Eric Idle and I were mates when we came to London and we did the odd show together. We did a bit of an obscure series called *Twice a Fortnight* which Michael Palin and Terry Jones wrote and did little films for.

People may not know what an accomplished musician you are. I've seen a video of you playing guitar with Mark Knopfler.

That doesn't mean I was playing it well! I have a blockage about rehearsing and practising, but I can mess around on certain instruments. Anybody hearing me might think I can play, but I can't really. I have a certain musicality, as I've written about 300 songs. The Who did a stage version of *Tommy* and they had guests playing various parts – I was playing the part of Cousin Kevin so I had a song and shared a mic with Davis Essex during the finale.

You worked on Jazz FM too.

I was there for two years – it was a dream, I had all Sunday morning to play anything I liked. It was called *Yes, but is it Jazz?* and the motive was to play a range of music which wasn't categorised.

In 2002 you were the subject of *This Is Your Life*, but you weren't very happy about that. Why?

When we worked together on *The Goodies* some years earlier, Tim Brooke-Taylor, Graeme Garden and I had agreed that if we were asked to do it, we wouldn't. So when Michael Aspel surprised me at Slimbridge, I refused. On the way back on the train my daughter Rosie phoned and said she hated me for not doing it – and she meant it. My wife also said it would be a bit 'awkward' if I refused (they'd got people from all over the country – which of course I didn't know). So through insult, aggression and threats, and by being made to feel such a heel, I gave in.

Is it true that fans of Reading Football Club sing 'Bill Oddie, Bill Oddie, rub your beard all over my body' to the tune of Madonna's song 'Erotica'?

Indeed, it is – when I found out I couldn't have been more amused and flattered!

What sorts of campaigning do you take part in?

I very much like Global Witness, an organisation which exposes international connections which are dodgy and gives them bad publicity. For example, I did a video for Global Witness at HSBC's headquarters, seeking answers to questions about why they were pumping large amounts of money into oil palm plantations – a huge problem in many parts of the world. We were physically ejected from the offices, which was the aim of the game. I'd always wanted to be doing an interview where a security man rushes over and puts his hand over the lens, and it happened!

You're also involved with the World Land Trust?

They're a terrific organisation, in many ways my favourite group, and they do achieve so much. It can't be bad when you've got Sir David Attenborough saying that this is as good a place as any that you can put your money. It's based on something I believe very strongly – and it's a practical rather than an ethical point – that you've got to own the land. And we've reached that point where just normal countryside doesn't exist any more. So the World Land Trust is based on the principle that any funds are to buy land, but then you must cooperate with the NGO on the ground. It's been very successful and provides a number of jobs for people.

Are you a supporter of the Green Party?

I'm a supporter of Caroline Lucas. I'm not a political person and don't really like being in that arena. I have been, but only when it's been directly connected to something I care about, for example the badger culls. I have had the opportunity to observe when various politicians are talking. Every now and again you hear someone you think you can trust, and Caroline Lucas falls into that category. She knows her stuff, she's a brilliant speaker, she's got a wry sense of humour, and I trust her.

You mentioned in a number of interviews that you have had a bit of depression over the years. Can you tell me a bit more?

I didn't know I had any problems until about ten years ago – I might have done before that but I wasn't aware of it. People would tell me I was moody and could be aggressive, and when I look back and reconstruct things I can see there was a sort of bipolar 'shape' in so far as I was very energetic and then I would get grumpy, but I'm lucky that it didn't go into dangerous elements. But when I was about sixty I had a really bad full-blown depression and I realised that there was something seriously wrong. It was really reprehensible of the people who were treating me, in that no-one ever mentioned bipolar and just said I had bad clinical depression. It was about nine or ten years till a consultant said everything he was hearing pointed to bipolar and put me on lithium. And it was wonderful – about two weeks later I was coming out of it and have been that way ever since. I just get the occasional winter blues with no extremes at all. I'm wary of being 'professional bipolar' but I did do a few talks at conferences for doctors and GPs to improve their attitude towards people who come to them.

Do you feel positive about the birding world and community currently?

The honest truth is that I'm nothing like as involved in birding as I used to be. I do join an occasional twitch, but when a Naumann's Thrush turned up at Chingford in 1990 I did not bother to go – even though it was just a few miles down the road. I'd rather be somewhere remote and find my own birds. The rare bird thing doesn't turn me on very much and I've no idea what my life or British lists are. I get as much pleasure out of doing an hour in the garden. One thing that has obviously changed is the technology. When I was young, if you had a telescope you must have been a real birdwatcher – but now everyone has one. Birdwatching also does seem to be turning quite rapidly into photography because of the ease of taking digital images.

If you could meet somebody from the history of ornithology, who would that be?

John James Audubon. Even two centuries after he created them his paintings are still fantastic.

Can you remember your best ever day's birding?

Central Park, New York, in spring.

And your worst day's birding?

I once went to Ireland and spent three days without seeing anything. It was really disappointing – but that's birding for you!

Where in the world would you like to go birding with all expenses paid?

I'd take a chance on hitting a good spring migration in Texas.

What new bird would you most like to see?

That's hard, because I don't really have target birds. I was delighted to see kiwis in New Zealand last year.

What is your favourite bird family?

Warblers, especially American ones.

What is your favourite bird book?

The House on the Shore by Eric Ennion. I was very fortunate to meet him several times at Monks' House on the Northumberland coast and he was an inspiration.

What is your favourite non-birding book?

Dead Famous by Ben Elton.

What is your favourite music?

I have several favourites: *Sometimes it snows in April* by Prince, *For free* by Joni Mitchell, *God's song* by Randy Newman, *I can't make you love me* by Bonnie Raitt, *Walking in Memphis* by Marc Cohn, and *Something so right* by Paul Simon.

What is your favourite film?

Field of Dreams.

What is your favourite TV programme?

Match of the Day – especially if my team, Manchester United, are playing.

LAST THOUGHTS

The seventeen birders interviewed here were chosen because of their interest in birds and/or birding, although some of them are known better for their contributions in other fields of human endeavour (most notably Frank Gardner, Tony Juniper, Ann Cleeves, Kevin Parr, Barbara Young and Bill Oddie). Like their predecessors in *Behind the Binoculars*, they were chosen to encompass a varied range of perspectives and approaches to birding, so their stories should complement each other – and we believe that they do.

Our interviewees share a couple of common characteristics, one an inevitable consequence of the other. The first is that as a group they are, let us say, mostly rather mature, with their birth years spanning from the 1930s (Bryan Bland and Tony Marr) to the 1970s (Kevin Parr and Dawn Balmer). As a consequence of this age distribution our interviewees are mostly men (thirteen men and four women), which is a comment on the past rather more than the present of ornithology and birding. We dipped back into a book published in the 1990s (*Who's Who in Ornithology*) and found that only around 5% of the entries were of women – which shows the scale of male dominance of this interest at that time. There has been some progress, and we are sure that a similar book in a couple of decades' time would have much better gender balance.

Most of the interviewees gained their interest in birds at a young age, before or during their teens, although one, Ann Cleeves, claims not to have an interest in birds to this day. In some cases, this interest was encouraged by parents, friends or teachers at school but in many cases it arose spontaneously and was maintained despite lack of companions with which to share it. Some were egg collectors in the days before it was illegal, when this was a common pastime of boys in particular. Some of our interviewees (Tim Birkhead, Roy Dennis, Tony Juniper) kept birds as pets in their youth.

Many of our interviewees played as children outdoors, either kicking a ball around in the street, park or a field, or roaming the countryside. That freedom will sound unusual to today's children and to their parents, and that sets a different context for children getting interested in nature these days. We feel that the freedom to roam and to explore was very important in introducing many of our interviewees to nature and cementing that strong relationship. Freedom to explore nature on one's own at an early age may have been more important than the lack of computer screens to distract them (after all, there were always books before computer games came along) in setting many ageing naturalists on their life paths.

City parks also played a part in several of our interviewees' initiation into nature watching, whether in north London, Leeds, Sheffield or Nottingham. It is unlikely that current-day visitors to those parks will find Hawfinches, Red-backed Shrikes, Lesser Spotted Woodpeckers or White-clawed Crayfish. Are they as rich in nature? Will they inspire today's young people? What might the nature memories of today's visitors be?

But time and again in these interviews people recall bird sightings from decades ago, and not always of rare birds but of common ones, which seem as fresh and clear as if they were made yesterday. The sights of birds have lived in these people's heads for most of their lifetimes.

If only our interviewees had all been equipped with mobile phones and the Birdtrack app in their youths, we could do a much better job of comparing their observations then with those possible now – and they would be quite different. It is clear that keeping a notebook of records was a habit that many of these young birders (including Richard Porter, Carol Inskipp and Tim Cleeves) got into and never lost, and there must be hundreds of their generation who have piles and piles of bird records that could be transferred to electronic format now and kept as a historical archive. The best of these (complete bird lists with known locations and dates) would be like ornithological gold dust. We wonder whether more effort should be put into capturing these data before they are thrown away and lost for ever.

But what would the data show, if we had them? There would be gains and losses, that's for sure, but Richard Porter's comment that one has to travel to places like Poland these days to see what the wildlife of the British countryside was like in his youth is probably not far short of the truth. The childhoods of the older interviewees were before the start of formal bird monitoring schemes in the UK. The only way to capture information from those times that otherwise might be lost would be to transfer records made at the time into some more lasting and collected source and to download the memories of those active at the time. We wonder which source would be the more accurate and the more illuminating – what is in the notebooks, or what is in the heads?

Travel was becoming a more and more feasible option as our interviewees grew up. Motorbikes and bicycles played a bigger part in their lives than they would these days, as did hitching lifts and breaking down in cheap cars. Our interviewees spanned a period when the world opened up for birdwatchers, with travel becoming easier and cheaper, and this helped produce an avalanche of bird guides to different countries and field guides to their birds. Searching for birds was the driving force behind travel for some, whilst others travelled through work and picked up lots of new bird sightings on the way. But it wasn't always easy, as Tim Inskipp's return journey from India via boat, arrests, begging, buses, taxis and trains demonstrates.

This book has stories from the high Arctic and Antarctic, and from the world's high mountains and deserts. And when asked where they would like to go birding our interviewees cover the globe with their wish lists. Everywhere has birds, and most places will have visits from foreign birdwatchers intent on seeking them out.

The world has over 10,000 species of bird, possibly nearer 11,000 now, and the number keeps going up as scientific knowledge leads to the splitting of existing known species into several new species. No-one on Earth has seen more of these than Jon Hornbuckle, who is well over 9,500 species and has probably seen more birds than all of the other interviewees in this book put together. But he hasn't seen a Slender-billed Curlew – a globally threatened species which may now actually be extinct despite once migrating across the Mediterranean area in huge flocks. Three of our interviewees (Tim Cleeves, Ann Cleeves and Richard Porter) mention encounters with Slender-bills in their interviews, but this bird is unlikely to form part of the recollections of future birders.

During the preparation of this book the world record for seeing the most birds in a calendar year has been broken twice: first by Noah Stryker (6,042 species in 2015) and then by Arjan Dwarshuis (6,841 species in 2016). Whatever you feel about the value of such exploits, they show that it is now feasible to see well over half of the world's birds in a single year if you have the time, the drive and the resources. In 1995 the world record was 3,662 species, which was only surpassed in 2008 by two of our interviewees in *Behind the Binoculars*, Alan Davies and Ruth Miller, with 4,327 species (and Ruth is still the female world record holder, of course).

Many of our interviewees have been involved with nature conservation and environmental protection, as volunteers, as staff, or as trustees of various nature conservation or environmental outfits. Not surprisingly, Tim Cleeves, Richard Porter, Tony Marr, Roy Dennis, Bill Oddie and Barbara Young all mention the RSPB in their thoughts and their comments.

RSPB membership only passed 10,000 in 1960 but now stands at over 1,200,000. Similar or even more impressive membership growth has been experienced by the range of conservation organisations with which our interviewees have been associated (and Barbara Young has been a trustee or council member of most of them). There has been a phenomenal increase in support, and in interest in wildlife in general and birds in particular. Whereas in the past the wildlife NGOs were, perhaps, formed of the enlightened end of the establishment, they can now be seen as popular movements – and that requires a different approach to interacting with government, industry and the media. A couple of interviewees reflect on whether the wildlife NGOs are striking the right balance between partnership and opposition in their relationship with government these days. These relationships evolve over time.

Politics inevitably comes into some of these interviews, as some of our interviewees (particularly Barbara Young and Tony Juniper) have been closely involved with trying to get a better deal for wildlife from the political system. Although the individuals involved in government are ever-changing, there are some interesting comments on the political characters of the fairly recent past in this volume. But as we head to print, we also head towards a general election in June 2017 after a change of Prime Minister in July 2016. It's always going to be a changing political world – in which the reliability of the natural world can be some comfort to us all.

The biggest political change in the UK for decades was initiated after these interviews were completed – the referendum vote to leave the European Union. We, like most others, hadn't seen this coming and didn't raise it as an issue with our interviewees, who have spent either all or most of their lives with the UK as part of the European Economic Community, European Community or European Union. The impacts for birders are unclear. There may be implications for travel, although we will probably not return quickly to the times that some of our interviewees will remember when European travel involved getting visas and travelling with a portfolio of different currencies. However, the departure of the UK from the EU will have implications for the conservation of all of our birds: UK conservationists will no longer be able to threaten UK governments with legal action in Europe if nature directives are not respected. There will also be implications, as in the EU at a greater political scale, for how much the UK is likely to be willing to invest financially in the workings of a geographic partnership of which it is no longer a member – will the RSPB and other UK organisations continue to invest at the same level in Birdlife International and similar international networks?

Birds bring very different people together – people of different ages, religions, politics and backgrounds. And there is no better place to meet that varied tribe of birders than at the annual Birdfair at Rutland Water, organised for so many years by Tim Appleton (and Martin Davies, and a team of helpers). Most of our interviewees have attended several Birdfairs, and like the thousands of others who go each year, what brings them together is a keen interest in birds. If our interviewees were to gather together at a future Birdfair then they might get on to the subjects of what books they have read, or politics, or a whole range of other subjects – but they would start with birds, and it is likely that the conversations about where they have been and what they have seen would keep this group of birders going for many an hour. Maybe we'll see them at Birdfair – and maybe we'll see you there too!

Selected Bibliography

Alain-Fournier. (1913) *Le Grand Meaulnes*. Émile-Paul.

Anon. (1969) *AA/Reader's Digest Book of British Birds*. Drive Publications.

Arnold, E.C. (1936) *Birds of Eastbourne*. Strange the Printer.

Arnold, E.C. (1940) *Bird Reserves*. H.F. & G. Witherby.

Balmer, D.E., Gillings, S., Caffrey, B.J., Swann, R.L., Downie, I.S. and Fuller, R.J. (2013) *Bird Atlas 2007–11: the Breeding and Wintering Birds of Britain and Ireland*. BTO Books.

Benson, S.V. (1937) *The Observer's Book of Birds*. Frederick Warne.

Benson, S.V. (1970) *Birds of Lebanon and the Jordan Area*. Frederick Warne.

Birkhead, T.R. (1991) *The Magpies: the Ecology and Behaviour of Black-Billed and Yellow-Billed Magpies*. A. & C. Black.

Birkhead, T.R. (1993) *Great Auk Islands: a Field Biologist in the Arctic*. A. & C. Black.

Birkhead, T.R. (2000) *Promiscuity: an Evolutionary History of Sperm Competition*. Harvard University Press.

Birkhead, T.R. (2002) *The Red Canary: the Story of the First Genetically Engineered Animal*. Weidenfeld & Nicholson.

Birkhead, T.R. (2008) *The Wisdom of Birds: An Illustrated History of Ornithology*. Bloomsbury.

Birkhead, T.R. (2012) *Bird Sense: What It's Like to Be a Bird*. Bloomsbury.

Birkhead, T.R., Wimpenny, J. and Montgomeriey, B. (2014) *Ten Thousand Birds: Ornithology since Darwin*. Princeton University Press.

Birkhead, T.R. (2016) *The Most Perfect Thing: the Inside (and Outside) of a Bird's Egg*. Bloomsbury.

Chang, J. (1991) *Wild Swans: Three Daughters of China*. Harper Collins.

Cleeves, A. (2006) *Raven Black*. Pan.

Cleeves, A. (2008) *White Nights*. Pan.

Cleeves, A. (2009) *Red Bones*. Pan.

Cleeves, A. (2010) *Blue Lightning*. Pan.

Cleeves, A. (2013) *Dead Water*. Pan.

Cleeves, A. (2014) *Thin Air*. Pan.

Cleeves, A. (2015) *Shetland*. Macmillan.

Cleeves, A. (2016) *Too Good To Be True*. Pan.

Cleeves, A. (2016) *Cold Earth*. Pan.

Cocker, M. and Inskipp, C. (1988) *A Himalayan Ornithologist*. Oxford University Press.

Cocker, M. and Mabey, R. (2005) *Birds Britannica*. Chatto & Windus.

Cramp, S. *et al.* (eds.) (1977–96) *Handbook of the Birds of Europe, the Middle East, and North Africa: The Birds of the Western Palearctic* (multiple volumes). Oxford University Press.

Delacour, J. (1954) *Waterfowl of the World* (4 volumes). Country Life.

di Lampedusa, G.T. (1958) *The Leopard*. Casa editrice Feltrinelli.

Elton, B. (2001) *Dead Famous*. Bantam Press.

Ennion, E. (1959) *The House on the Shore*. Routledge & Kegan Paul.

Evans, G. (1954) *The Observer's Book of Birds' Eggs*. Frederick Warne.

Ferguson-Lees, J., Willis, I. and Sharrock, J.T.R. (1983) *The Shell Guide to the Birds of Britain and Ireland*. Michael Joseph.

Fitter, R.S.R. and Richardson, R.A. (1952) *The Pocket Guide to British Birds*. William Collins.

Fleming, R. L. Sr and Fleming, R. L. Jr (1976) *Birds of Nepal with Reference to Kashmir and Sikkim*. R. L. Fleming Sr and Jr.

Gibran, K. (1923) *The Prophet*. Alfred A. Knopf.

Greene, G. (1951) *The Third Man*. Viking.

Grimmett, R., Inskipp, C. and Inskipp, T. (1988) *Birds of the Indian Subcontinent*. Christopher Helm.

Grimmett, R., Inskipp, C. and Inskipp, T. (2000) *Birds of Nepal*. Christopher Helm.

Hardy, T. (1874) *Far from the Madding Crowd*. Cornhill Magazine.

Harrison, P. (1983) *Seabirds*. Croom Helm.

Hayman, P. (1979) *The Mitchell Beazley Birdwatcher's Pocket Guide*. Mitchell Beazley.

Inskipp, C. (1988) *A Birdwatcher's Guide to Nepal*. Prion.

Inskipp, C. (1989) *Nepal's Forest Birds: their Status and Conservation*. International Council for Bird Preservation.

Inskipp, C. and Inskipp, T. (1985) *A Guide to the Birds of Nepal*. Croom Helm.

Inskipp, T. (1975) *All Heaven in a Rage*. RSPB.

Inskipp, T. and Thomas, G.J. (1976) *Airborne Birds: a Further Study into the Importation of Birds into the United Kingdom*. RSPB.

Jonsson, L. (1992) *Birds of Europe with North Africa and the Middle East*. Christopher Helm.

Juniper, T. and Parr, M. (1998) *Parrots: a Guide to the Parrots of the World*. Yale University Press/Pica.

Juniper, T. (2003) *Spix's Macaw: the Race to Save the World's Rarest Bird*. Fourth Estate/Atria.

Juniper, T. (2007) *How Many Lightbulbs Does it Take to Change a Planet? 95 Ways to Save Planet Earth*. Quercus.

Juniper, T. (2007) *Saving Planet Earth*. Harper Collins.

Juniper, T. (2013) *What Has Nature Ever Done For Us?* Profile.

Lansing, A. (1959) *Endurance: Shackleton's Incredible Voyage.* Carroll and Graf

Leopold, A. (1949) *A Sand County Almanac.* Oxford University Press.

Lloyd, G. and Lloyd, D. (1971) *Birds of Prey.* Bantam Books.

Mullarney, K., Svensson, L., Zetterström, D. and Grant, P. J. (2009) *Collins Bird Guide*, 2nd edition. Harper Collins.

Nethersole-Thompson, D. and M. (1979) *Greenshanks.* T. & A. D. Poyser.

Oddie, W. (1980) *Bill Oddie's Little Black Bird Book.* Eyre Methuen.

Parr, K. (2014) *The Idle Angler.* Medlar Press.

Parr, K. (2014) *The Twitch: Birdwatching Can be Murder.* Unbound.

Parr, K. (2016) *Rivers Run: An Angler's Journey from Source to Sea.* Ebury.

Pemberton, J.E. (1997) *Who's Who in Ornithology.* Buckingham Press.

Peterson, R.T. and Fisher, J. (1956) *Wild America.* William Collins.

Peterson, R.T., Mountfort, G. and Hollom, P.A.D. (1954) *A Field Guide to the Birds of Britain and Europe.* William Collins.

Porter, R. and Aspinall, S. (2010) *Birds of the Middle East*, 2nd edition. Christopher Helm.

Porter, R. and Aspinall, S. (2017) *Birds of the Middle East*, Arabic edition. Christopher Helm.

Pye-Smith, E. (1946) *The Birds' Alphabet Book.* C. & J. Temple.

Rackham, O. (1994) *The Illustrated History of the Countryside.* Weidenfeld & Nicolson.

Ransome, A. (1947) *Great Northern?* Jonathan Cape.

Sagar Baral, H. and Inskipp, C. (2004) *The State of Nepal's Birds.* Bird Conservation Nepal.

Sagar Baral, H. and Inskipp, C. (2005) *The Important Bird Areas of Nepal*. Bird Conservation Nepal.

Salinger, J.D. (1951) *The Catcher in the Rye*. Little, Brown and Company.

Sandars, E. (1933) *A Bird Book for the Pocket*. Oxford University Press.

Shackleton, E. (1919) *South: The Story of Shackleton's Last Expedition 1914–1917*. William Heinemann

Simpson, J. and Weiner, E. (1989) *The Compact Oxford English Dictionary*, 2nd edition. Oxford University Press.

Smith, S. (1946) *How to Study Birds*. William Collins.

Steinbeck, J. (1952) *East of Eden*. The Viking Press.

Tolkien, J.R.R. (1954) *The Lord of the Rings*. George Allen and Unwin.

Tolstoy, L. (1869) *War and Peace*. The Russian Messenger.

Vesey-Fitzgerald, B. (1956) *A Third Book of British Birds and their Nests*. Ladybird Books.

Wales, the Prince of, Juniper, T. and Skelly, I. (2010) *Harmony: a New Way of Looking at Our World*. Blue Door.

Wales, the Prince of, Juniper, T. and Shuckburgh, E. (2017) *Climate Change*. Ladybird Books.

Webb, C. (1963) *The Graduate*. New American Library

Witherby, H.F., Jourdain, F.C.R., Ticehurst, N.F. and Tucker, B.W. (1938–41). *The Handbook of British Birds*. H.F. & G. Witherby.

Index

Mullarney, Killian 159
Murrelet, Ancient 85
Murrelet, Long-billed 84
Muscat, Oman 91
Myanmar 67, 158
Mynott, Jeremy 80

naming and shaming 111
National College of Food
  Technology 123
National Editorial and
  Writing Services 136–7
National Health Service
  (NHS) 163–6, 167–8,
  170
Nature Conservancy 26,
  28, 130
Nature Conservancy
  Council 53
Naturetrek 89, 104
Needs Ore 24
Nene 63
Nepal 151–2, 153, 154,
  155–8, 181
*Nepal's Forest Birds: their
  Status and Conservation*
  151
nest protection schemes 15,
  26, 28, 30–1, 96–7
Nestlé 118
Netherlands 27, 154
Nethersole-Thompson,
  Desmond 92
Nethersole-Thompson,
  Maimie 92
New Forest, Hampshire 23,
  24, 45
New Guinea 158; *see also*
  Papua New Guinea
New Networks for Nature
  80
New York 97, 178, 189
New Zealand 8, 10, 154,
  189
Newbiggin, Northumber-
  land 16
Newcastle University 72
News of the World 165
Newton, Ian 76, 78
Nicholson, Carl 16
Nicholson, Max 26
Nicol, Wendy 170
Nigeria 98
Nightingale 37

Nightingale, Thrush 180
Nightjar 89
Nisbet, Ian 124
Noe, River 96
non-governmental organi-
  sations (NGOs) 172–3,
  194–5
  activism 117–18
  business partnerships
  118–19
  Middle East 130
  Stop Climate Chaos 112
  *see also individual
  organisations*
Norfolk 58, 84, 85, 92, 137,
  140–1, 143–4, 145
  Blakeney Point 120, 132
  Broads 147
  Cley 70–1, 92, 120,
  122–3, 126, 138, 149
  Hickling Broad 147
  Salthouse Broad 59
  Thetford Forest 89
North, June 83
North Rona, Outer
  Hebrides 26
North Sea oil industry 27,
  28
Northumberland 16–17,
  20, 147
Northwest Passage, Canada
  57, 60
Norway 7
notebooks 3, 12–13, 122,
  150, 177, 192–3
Nottingham 64, 93, 150
Nottingham High School
  93

OBE (Order of the British
  Empire) 182–3
Observer's Book of Birds 14,
  47, 108, 133, 135, 149
*Observer's Book of Birds'
  Eggs* 108
Oddie, Bill 176–90
Oddie in Paradise 181
oil palm plantations 187
oil spills 30, 127–8
Old Scone, Perthshire
  160–2
Oman 10, 90–1, 129
Oriole, Baltimore 25
Oriole, Golden 2

Ornithological Society of
  the Middle East (OSME)
  129
Osborne, George 113, 114
Osprey 26, 29, 31, 32, 147
Otago Peninsula, New
  Zealand 8
Otter 42–3
Ouse Washes, Cam-
  bridgeshire 178
Out Skerries, Shetland 178,
  180
Outer Hebrides 26, 57,
  58–9
overseas birding 193–4
  Bill Oddie 180–1
  Bryan Bland 140
  Carol Inskipp 151, 153,
  154, 155–6, 157
  Dawn Balmer 89, 90–1
  Frank Gardner 1–2, 4–9
  Jon Hornbuckle 97–105
  Richard Porter 124, 129,
  130, 131, 180–1
  Roy Dennis 31–2
  Tim Appleton 66–7
  Tim Inskipp 151, 152,
  153–7
  Tony Marr 56–8
Overseas Territories, UK
  130–1
Owl, African Eagle 5
Owl, Long-eared 151
Owl, Short-eared 151
Owl, Snowy 10, 38, 180
Owl, Socotra Scops 5
Owlet, Long-whiskered 104–5
owls 60, 158
Oxford University 94
Oxfordshire 106
Oystercatcher 82
Oystercatcher, American 58

Packham, Chris 105, 117
Pagham Harbour, West
  Sussex 50–1, 56
Painted-snipe, Greater 142
Pakistan 6, 97, 98, 100, 153
Palin, Michael 105, 186
Palmyra, Syria 4
Panama 154
Panda, Giant 32
Papua New Guinea 6–7, 99,
  101, 143, 181